The Spatial Edge

Thomas Fisher and Len Kne

The Spatial Edge

The Strategic Advantage of GIS Skills Across Higher Education

Esri Press

Redlands, California

Contents

CONTENTS BY Discipline

The contents of this book are divided into four categories of interest: contents by discipline, by industry, by organization, and by location. These categories should help you locate the topic or topics that best meet your field of interest.

Public Health

Science and Technology

CONTENTS BY Industry

CONTENTS BY Organization

Preface

Universities consist of faculty, staff, and students working together to move knowledge forward, and so we thought that, to move this book forward to completion, we needed an equally diverse team, with a faculty member, staff member, and graduate student working together. The three of us—Tom Fisher, Len Kne, and Ankila Kumari, respectively—have spent a lot of time making this handbook as useful, informative, and engaging as possible, and we hope that you find it so.

Before you take our word for it, though, we thought we should map the journey we've been on and tell a bit more about ourselves in suggesting why you might want to read this handbook. This book arose out of a conversation that Tom Fisher had with Esri's president and cofounder, Jack Dangermond, and some of his Esri® and higher education colleagues on the patio off the lobby of Esri's headquarters building in Redlands, California. They saw the need for a book that documented the impact that geographic information systems (GIS) was having in higher education, not just among obvious spatial fields such as geography or planning but in disciplines that are rarely thought of as GIS related. They also envisioned a book that helped those interested in using GIS in their teaching and research to access the technology more easily and use it more effectively. With that conversation, this book began.

Tom immediately enlisted his GIS colleague, Len Kne, in the project, knowing that Len, who directs the U-Spatial unit at the University of Minnesota, knew the most about the disciplines using GIS, not only at Minnesota but in other institutions around the world. With support from Esri, Tom and Len hired an energetic and enthusiastic graduate student in the master of geospatial information science (MGIS) program at Minnesota to be their research assistant. Each of us took a circuitous path getting here; by recounting our journeys, we thought it might be useful for others on their own journeys to a spatial understanding of the world.

Tom's journey

Tom is the oldest of the three of us and so his journey has been the longest. Like many young people in the American Midwest, he grew up getting to know the larger world through *National Geographic Magazine*, through its articles and its extraordinary maps, which he hoarded to the displeasure of his parents, who rarely saw them again once the issues made it to his room. And while other students daydreamed about sports or fashion, Tom would doodle imaginary maps in the borders of notebook pages, with roads winding their way through hills and valleys to towns sitting next to coastlines.

Tom went to architecture school with the intention of working in his family's architectural firm in Detroit, but in school, he met Lewis Mumford, the noted historian of cities, and studied with Colin Rowe, a well-known urban designer. They rekindled his youthful interest in maps and mapping not just the physical features of places but their social, economic, and cultural characteristics. Tom never did join his family's firm. He went to two graduate schools and, after receiving his degrees, became a writer, editor, and eventually a professor and dean, teaching urban design and mapping to thousands of students over the years.

In addition to teaching, he now directs an endowed center at the University of Minnesota that uses geospatial tools, such as GIS, to help communities of diverse sizes and compositions map their assets and leverage those assets to imagine better futures for themselves, to reduce their impacts on the natural environment, and to build new communities around shared interests and ecologies. Maps help connect people, engage them, and build consensus where none seemed to exist.

In so many communities, especially those that have lost population or long been underserved or overlooked, residents often focus on what their communities once had that they've lost. Maps help move people past a deficit mindset based on loss to an abundance mindset addressing what they still have to work with and build on. In their community work, Tom and his team often use static maps that allow residents to mark them up and digital maps that let them see possibilities and understand the consequences of their ideas. Maps make all the difference in building consensus and helping communities remember why they love the place in which they live. It's gratifying work.

Len's journey

Len's path into academia also looked nothing like that of most of his colleagues, who followed the well-trodden route of undergraduate degree, graduate degree, and often a PhD. Instead, Len zigzagged through multiple college dropouts, assorted adventures, and jobs that slowly nudged him toward a mapping career. Around 1987, he left school to join Hennepin Parks—now known as Three Rivers Park District—in Plymouth, Minnesota, as a research specialist, spending the next seven years doing marketing research, talking to park visitors, conducting needs assessments, and tracking how people use park spaces.

One of his regular tasks was mapping park use before GIS and smartphones were widely available, so his mapping "system" was a clipboard, pencil, and paper map, marking where picnic groups were sitting, how many people were in each group, and what they were doing. Those maps went into filing cabinets, and when analysis was needed, someone dug through the folders and manually tallied the results. It worked, but it always left him thinking that there had to be a better way. Having the first personal computer at the park district sparked an interest in learning more about how to manage, store, analyze, and visualize data (sounds like GIS, right?). Soon he was deep into information technology and on to a series of jobs, helping companies and government agencies install networks, manage databases, develop apps, and figure out how technology can solve real-world problems.

In 2004—more than two decades after graduating from high school—he finished his undergraduate degree in recreation resource management. One of the last classes he took was Introduction to GIS with Paul Bolstad, a name many in the field will recognize. Paul literally wrote the book, *GIS Fundamentals*—one of the seminal GIS textbooks still in use around the world. For the first time, Len saw how his IT skills could intersect with geography. He realized that the piles of paper maps he once filed away could live in a GIS, ready for analysis. He was hooked.

In 2009, Len completed an MGIS degree at the University of Minnesota, which opened new doors for him: consulting gigs, a stint with the Metropolitan Council, and finally the opportunity to return to the University when U-Spatial was founded in 2011. He was employee number two, right after Kris Johnson, the first grad student hired. From there, Steve Manson, Francis Harvey, and many others built something bigger than he could have imagined. He has been at U-Spatial for more than 13 years, far longer than he had spent at any job, because, every day, he gets to learn and be challenged by so many amazing colleagues.

Ankila's journey

Ankila's journey into the world of geographic information science was anything but planned. She grew up in a small village in Bihar, India, a place where maps were printed on paper, not displayed interactively on a computer screen, and where careers in technology usually meant engineering, medicine, or government service. GIS was not even a term she had heard growing up; it wasn't even on her radar until much later in life, long after she had completed a degree in a completely different field: automobile engineering at IEC College of Engineering and Technology in Greater Noida. At the time, she was immersed in engines, mechanical systems, and design blueprints, and she learned technologies such as computer-aided design (CAD), computer-aided manufacturing (CAM), and AutoCAD, envisioning a future in the automobile industry, designing cars or working in manufacturing plants. Her career path seemed set to be practical, mechanical, and hands-on.

After graduation, she began the job hunt, sending out résumés and hoping for that one break. One day, a friend forwarded her an email mentioning a hiring opportunity at Cyient for AutoCAD engineers. It sounded like a perfect fit, so she applied immediately. But when she arrived for the interview a week later, the company was hiring junior GIS engineers. She had never heard of GIS; sitting outside the interview room, she pulled out her phone and typed "What is GIS?" That quick search changed everything. She began to see GIS as more than just maps; it was about spatial data, location-based analysis, and solving real-world problems. That moment sparked a curiosity in her that would reshape her life.

She started at Cyient as a junior GIS engineer and underwent about 15 days of training before being assigned to projects. She threw herself into the work, not just learning software but trying to understand the logic and applications of GIS, ending up as the top trainee in her group. At Cyient, she didn't just work, she thrived. Ankila learned to optimize workflows, explore new tools, and share ideas. GIS became more than a job—it became a way of thinking, of understanding patterns, of solving problems through spatial relationships.

Years later, she chose to deepen her knowledge and joined the MGIS program at the University of Minnesota, learning what many call a "spatial turn," in disciplines beyond geography, such as public health, history, literature, environmental science, and business, incorporating spatial data and mapping tools into the work. But this shift isn't happening everywhere. In India, where she completed her undergraduate studies, GIS is still largely confined to geography departments. There were no lectures, workshops, electives, or discussions of geospatial tools, and many students, like her, could find passion and purpose in GIS if they knew it existed.

Our journey to this book

Because all three had unconventional and multidisciplinary paths to writing this book, we were drawn to writing a book about colleagues around the world using GIS to explore everything from health disparities and migration trends to environmental changes and historical mapping, approaches that enrich learning and open up new ways to understand the world. A process often used when working with communities is *appreciative inquiry*, in which we get people to discuss what they love about what they do and what issues concern them, often discovering that others, in other fields or other places, share interests or areas of focus. Maps facilitate such inquiries.

Whether it's through the Minnesota Design Center that Tom directs or through U-Spatial that Len directs and where Ankila works, we find that data-rich, digital maps—GIS—link people and places in ways that many haven't thought about. The power of GIS lies in the juxtaposition of information, in the form of spatial layers, that reveals relationships that may be invisible even to those who have worked in a field or lived in a place for a long time. Maps help make meaning in the world.

GIS does more than merely document what was and what is—it reveals what could be, allowing us to imagine what doesn't yet exist to address present needs and past wrongs. Tom has been part of a movement called *geodesign*, which uses maps to project into the future, analyzing the challenges and opportunities of a situation, mapping possible scenarios and solutions to address them, and building a consensus around the responses that seem most promising and effective. Geodesign applies the methods of designers at a geographic scale, and although every designer does this slightly differently, the basic elements are the same.

Geodesigners start by understanding the existing context and the nature of challenges in that context and looking for precedents or analogies for how others have addressed those challenges, across time and in a variety of places. Based on that research, geodesigners develop alternative ideas—often as many as possible—which they analyze according to project goals until the options that meet the greatest number of goals in the most effective and acceptable way emerge. In this way, geodesigners eventually arrive at the best possible solution, which might be one alternative or a combination of several.

This process used to be top-down, with design teams doing this work mostly apart from the communities most affected by their decisions. That has largely changed, in part because of the often-disastrous consequences of designing in a vacuum. Over the last 50 years, design has become a much more inclusive and participatory process in which the voices of community members and the diversity of perspectives play a central role, especially in the understanding of context, the generation of ideas, and the assessment of options.

Design and geography are linked in their spatial way of seeing the world, whether it's the floor plan of the spaces in a building, the site plan of the spaces in a landscape, the street plan of the spaces in a city, or the flow plan of ecosystems or infrastructure in a region. Design and geography have an abiding interest in a range of fields, because almost everything eventually comes into being in some place, somewhere—an idea that underlies this book. Every field has a spatial or geographic component, just as it has a temporal or historical record. And a lack of awareness remains a major challenge.

Many people see GIS as a niche technical skill rather than a versatile tool that allows us to see the world in new ways. Biology students may study ecosystems without mapping them, and literature students might discuss settings without visualizing them spatially. These missed connections represent lost opportunities for learning, innovation, and impact. You don't need a geography background to begin—you just need curiosity, persistence, and someone to point you in the right direction.

Which is why we believe GIS should be introduced more broadly in higher education. Exposure shouldn't depend on luck, as happened for the three of us. Students deserve a chance to explore spatial thinking early on and see how it intersects with their interests, whether that's agriculture, urban development, public health, or the arts. This book stems from that belief. We have gathered examples of how GIS is being integrated into unexpected corners of academia: English departments mapping literature, chemistry labs tracking pollution, and business schools using location analytics. These examples prove that spatial thinking is not confined to geography—it has a place in every discipline.

Through this book, we hope to spark awareness not just among students but among educators, administrators, and policymakers. GIS is not merely a technical skill; it's a way of seeing connections, asking new questions, and finding solutions that matter. And maybe, just maybe, someone reading this will have their own "Google search moment," as Ankila did that day outside the interview room. Because sometimes, one unexpected question—"What is GIS?"—is all it takes to change everything. Just as each of us found GIS in unexpected ways, we believe universities and their communities can, too, sparking new connections, new disciplines, and new possibilities.

The following map shows the locations of all examples found in this book.

Acknowledgments

The speed with which this book came together is remarkable. In less than a year, what began as an idea on an Esri patio in Redlands, California, became a finished manuscript, and we're amazed at how thoughtfully and seamlessly it all came together. This wouldn't have been possible without the contributions of many generous and committed individuals.

We are especially grateful to the team at Esri for their partnership and encouragement throughout the process. Special thanks go to Jack Dangermond. This book was truly his idea, and through his vision, support, and direction, we were able to bring it to life. We are equally thankful to the team at Esri Education, including Angela Lee and Michael Gould, who helped us make key connections and identify resources that strengthened the book in meaningful ways. The team at Esri Press included Katie Gezi, Victoria Roberts, David Oberman, Carolyn Schatz, Christian Harder, and Stacy Krieg, whose guidance and professionalism (and nudging) kept this project moving forward.

At the University of Minnesota, we're grateful to those who helped build the foundation for the work described in these pages. In particular, we thank Steve Manson, Francis Harvey, and Claudia Neuhauser, whose early vision for U-Spatial recognized the importance of geospatial thinking across an entire institution. All the people who have worked with U-Spatial have been amazing, as well as our GeoForce colleagues and GIS professionals working across institutions who have taught us so much. Finally, we offer our deepest thanks to Ankila Kumari. She was drawn into this project with little warning and perhaps little sense of what she was signing up for, yet her dedication and thoughtful contributions were constant throughout.

CHAPTER 1

Democratizing spatial thinking across higher education

Thinking spatially brings people together around a common understanding of place, of community, of *where*. Like reading and math, GIS is a tool we use to enable spatial thinking, a means to an end. Reading isn't the reward—reading is what enables you to learn and view new experiences, other opinions, and diverse cultures. Math isn't the end—it allows you to model, measure, and look at the world in different ways. When we're trying to promote spatial thinking at a university, or in any discipline or sector, it's not about the act of thinking spatially; what's important is what we can learn while using this knowledge and GIS tools and looking through a spatial lens.

Thinking spatially is not just for geographers; building support for GIS democratizes spatial thinking across every discipline, and allows students, faculty, and staff to use the power of where to change the world. This book doesn't get into the pedagogy of teaching formal GIS courses; many other resources provide that information. Rather, the goal of this book is to show more than 100 examples of how people from around the world are using GIS in their teaching, learning, and research—and how building a geospatial support center (or services) at your institution is critical for delivering GIS in classrooms, research, outreach, and the operations of educational institutions.

But thinking spatially is just the beginning. The power lies in how you'll make the world a better place with these tools. If you think of departments and disciplines at a university as silos (which they often are), GIS can be a bridge that connects people and ideas across the university. That's the key ingredient in the recipe for spreading spatial thinking more widely and making GIS a part of teaching and learning, research and discovery, outreach and public service. Many people talk about the importance of interdisciplinary work, about mixing the diverse ingredients of colleges and universities into new dishes, but the structure of most institutions gets in the way. Geospatial tools offer an answer because even though they cannot restructure higher education, they can remap it, connecting disparate disciplines in new ways around shared questions or problems. Showing what's possible through the examples in this book will, we hope, give you what you need to convince your colleagues of the value of GIS and, in business terms, provide a huge return on investment.

Some universities are taking a holistic approach to integrating GIS across disciplines, such as environmental science, business, public health, and social studies. This approach includes teaching geography fundamentals, such as scale, projection, and Tobler's First Law of Geography, when working with geospatial data. As part of a broad approach, we believe that higher education should emphasize project-based learning opportunities and internships for students who need hands-on experience with spatial thinking to thrive in today's data-driven world. This approach prepares students with both technical GIS skills and critical thinking and problem-solving abilities. Furthermore, academia should partner with industry to help shape the curriculum, ensuring that students graduate with skills directly aligned with workforce needs. By blending academic training with real-world experience, these efforts make graduates highly effective in fields that rely on location intelligence.

The technology and tools for expanding spatial thinking across the entire campus have never been easier. For example, ArcGIS® Online has become a widely adopted platform for geospatial data analysis, offering accessible tools for mapping, spatial analysis, and real-time collaboration. Introduced in 2012, it has been a game changer in how we all map and perform location analytics. As a cloud-based GIS solution, it enables anyone to create, share, and analyze maps and data, making it invaluable for fostering spatial fluency among students, faculty, and staff—and the future workforce. Anyone can use this web-based tool without specialty software and with reasonable training. By providing hands-on experience with industry-standard tools, ArcGIS Online equips people to handle geospatial data effectively, which aligns with the demands of a modern, geospatially skilled workforce.

U-Spatial

As an example of a holistic approach to GIS, we want to share our story at the University of Minnesota with building U-Spatial, an internationally recognized geospatial support center. For context, the University of Minnesota is a large land grant university spread over five campuses, with 68,000 students from 130 countries and 25,000 faculty and staff. It's one of the more comprehensive universities in terms of disciplines represented and has $1.4 billion in annual research expenditures and more than 240 start-ups created in the last decade.

U-Spatial at the University of Minnesota has grown into a successful center with the mission to serve and drive the ever-present, fast-growing need across the university for expertise, data, and tools related to GIS, remote sensing, and spatial computing. Although U-Spatial is successful, that doesn't mean it's necessarily the right model for your university; you can find alternative approaches in chapter 7 ("Building a Geospatial Support Center").

A week in the life of U-Spatial

The breadth and depth of GIS across all disciplines and departments is inspiring. Every week, the team at U-Spatial lends its expertise to colleagues who are asking questions, looking for data, creating models, and developing apps to share their discoveries. We track the pulse of these interactions every Monday morning when Slackbot (a Slack tool that helps users perform work-related tasks) reminds us to share what we're working on. What follows illustrates some of what happens at U-Spatial in a week.

Our team in the Duluth, Minnesota, office has cornered the market in the state for creating hazard mitigation plans for counties and tribal nations to meet Federal Emergency Management Agency (FEMA) requirements. This work heavily relies on GIS and shares the information through ArcGIS Hub℠ sites.

Another team is deep in a National Institutes of Health (NIH)-funded project developing the Healthy Communities data portal that allows researchers to join their data to a growing array of data on socioeconomic status, ethnicity and race, and mortality. This effort allows colleagues to focus on their research rather than harmonizing data.

Team members Jess and Olena are working on a project looking at air pollution and mortality risk in veterans with chronic respiratory disease. All they have to do is geocode 12 million patient records and spatially join the records with risk factors and outcomes.

The online National Zoning Atlas has the audacious goal of conflating thousands of zoning maps and regulations across the United States. This week, Liz is working on two cities in Minnesota, Redwood Falls and Sleepy Eye.

Liz is also monitoring the help desk this week and helping a student with ArcGIS Experience Builder.

Shaam is working with ArcGIS Knowledge to map and graph how the 340-plus centers and institutes at the university interact with the United Nations' Sustainable Development Goals (SDGs) and one another.

Research into turfgrass (the grass on a golf course or sports field) is fascinating. Really. A group of researchers is looking at what causes winter turf damage, with developing new cultivars (cultivated plants) and management practices to reduce inputs such as fertilizer and water.

A couple of people are administering our ArcGIS Online subscription. With 20,000 active users and more than 150,000 maps, apps, and datasets, there are always management tasks to keep up on, although Python scripts handle nearly all that work.

Olena is collaborating with faculty on disparity ratio analysis using Clean Water Act violations data for the Mississippi River corridor.

More turfgrass: Several people are working on a USDA proposal looking for a better way to normalize the rating of test plots.

Another help desk ticket: help rebuild a rhizomes institutional application for ArcGIS Experience Builder.

Dena is working on a model of the cost of invasive species for the Minnesota Pollution Control Agency.

Here's a tough one: using satellite imagery to detect and monitor change of *Phragmites*, an invasive wetland grass, in Minnesota.

Marylee created a statewide map that shows all 1.5 million water connections to public water systems in Minnesota to identify whether the pipes contain lead. This ongoing project allows Minnesota residents to see whether they have lead waterlines and provides policymakers with data to plan how to remove all the lead lines.

The Spanish 101 classes are using ArcGIS Online for an assignment this week, so we'll see a surge of new people signing on for the first time.

The list could go on and on. We help a lot of people with that staple of GIS: geocoding. Geocoding is magic to someone who hasn't seen it before. Being able to map where grocery stores are located or where COVID-19 cases are spiking gives our colleagues new ways of seeing their data and sparks ideas for solving problems. At U-Spatial, we get to see people light up with ideas all the time.

The early years (2011–2015)

U-Spatial was started in 2011, when the Office of the Vice President for Research awarded a group of faculty at the university an Infrastructure Investment Initiative (I^3) grant totaling $2.5 million. The I^3 grant is a highly competitive program that usually funds equipment—think big microscopes and functional MRIs. U-Spatial was unique in that the faculty group from 16 departments and centers made the case that knowledge and geospatial expertise are infrastructure, and that they would benefit the entire university. This group, led by Steve Manson and Francis Harvey, both in geography, had spent a decade before that pulling together faculty, students, and staff from across the university through meetings, colloquiums, and seminars to craft the message that resulted in this initial funding.

An *ArcNews* article from 2013, "U-Spatial: A Consortium for the Spatial University" by then-director Francis Harvey, outlined the approach of the new center. Originally, U-Spatial had four distinct focus areas called cores: analysis, imaging, data, and the central core.

Analysis core

Research on complex systems and issues such as climate variability, public health, and rapid social change require advanced spatial analysis and coordinated geospatial infrastructure. The analysis core focused on using interdisciplinary research on human-environment systems to establish a durable foundation for integrated spatial research. During this period, the analysis core made good headway in defining models for a collaborative geodesign environment that supported applied research and decision-making in public affairs, design, and environmental and natural resource disciplines.

The analysis core also advanced complementary efforts in modeling and mapping through collaborations among multiple centers at the university, including the Institute on the Environment, Computer Science, and Engineering and the Center for Urban and Regional Affairs. Together, these efforts reinforced three tightly connected areas of emphasis: modeling, geodesign, and mapping. Modeling activities focused on reusable computational frameworks and high-demand datasets, including spatially enabled public health data linked to census information and parcel-level data. Geodesign initiatives emphasized collaborative, technology-enabled approaches that integrated spatiotemporal modeling and 3D visualization to support scenario evaluation and decision-making. Mapping efforts built on the early adoption of cloud-based GIS platforms and longstanding GIS and web-mapping programs, supported by extensive historical and contemporary spatial data holdings curated by research libraries. Collectively, these activities strengthened analytic capacity and supported research across domains ranging from social equity to ecosystem services.

Imaging core

Remote imaging is essential for studying environmental and urban change, but the volume of data and technical expertise required can limit students' and researchers' ability to fully use these resources. The imaging core addressed these challenges by supporting research at all scales, making remote imagery more accessible to colleagues and building on expertise and systems.

This work continues today as part of U-Spatial. For example, U-Spatial sponsors a campus agreement with Planet Imagery that provides global imagery to researchers and learners. Providing access to the data is just the start; resources and expertise are available to create workflows to process the imagery through a variety of platforms, including ArcGIS Pro and the high-capacity computing resources at the Minnesota Supercomputing Center. These workflows are adapted for other data resources (for example, Landsat, Sentinel, lidar point clouds, and so on) as needed.

Data core

The data core focused on helping researchers cope with the flood of large, complex datasets by making spatial data easier to archive, manage, discover, and reuse. Working with the university libraries, the Minnesota Population Center (also known as IPUMS, an organization that provides census and survey data), and campus information technology partners, the data core explored open, standards-based data architectures to support data sharing within the university and beyond, developed early tools for web-based spatial data discovery, and addressed long-term preservation needs.

This work also led to the development of shared server infrastructure to support repositories, indexing, and spatial data services. In practical terms, individual centers or departments no longer needed to invest in the expertise to procure and maintain ArcGIS Enterprise servers or spatial databases. U-Spatial provides this infrastructure at no cost to anyone at the university, allowing for consistency in resources and ensuring that best practices for data security are followed.

Central core

The central core consolidated the support of GIS into one location. This consolidation included a help desk where any question could be asked. Early on, the most common question was: How do I install ArcGIS Desktop on my computer? That changed with rapid adoption of ArcGIS Online and ArcGIS Pro, and now most people can access these tools easily on their own. The help desk support makes it easy to find answers that were previously unearthed if you could find your own GIS geek in your corner of the campus. The help desk can answer more than 90 percent of the questions it receives. When inquiries cannot be answered, the help desk can reach out to the GIS community for assistance. This approach works well because experts across the university no longer need to answer routine questions but can build collaborations with colleagues when their expertise is needed.

The central core had three other primary functions: training workshops, consulting, and building a GIS community through events and communications. Not every student has the time to take one or more semester courses, so workshops are key to spreading spatial fluency across the university. The workshops focus on introducing people to GIS and story maps (a form of digital storytelling with maps and multimedia using the ArcGIS StoryMaps[SM] app), as well as specific areas of interest, such as data collection. These workshops are designed to complement individual learning opportunities, such as self-paced ArcGIS tutorials and e-learning in Esri Academy. Still, we see a market and need to provide synchronized learning opportunities, in person and online for people who prefer to learn with that approach. Stitching workshops together provides a further opportunity to implement microcredentials, such as Credly badges (digital credentials). We have a popular badge on Digital Storytelling, which is completed by taking Webmapping 101, StoryMaps 101, and then creating a story map.

Joining Research Computing (2015–2019)

Reaching the end of the initial five-year grant for U-Spatial was a stressful time. There was no certainty about how the unit would achieve long-term sustainability. We had been notably successful in generating significant geospatial research across the four cores—analysis, imaging, data, and central—and fundamentally shifting how geospatial support functioned at the university by establishing a help desk, consulting services, and a range of training opportunities and workshops.

Then, a sustainable solution came together in a way that exceeded our expectations. As the original funding was ending, a new unit was being formed within the Office of the Vice President for Research: Research Computing. At that time, Research Computing primarily consisted of the Minnesota Supercomputing Institute and the

University of Minnesota Informatics Institute, which was still in the early stages of defining itself within the emerging field of data science.

We were fortunate that then-director of Research Computing Claudia Neuhauser recognized the importance of geospatial work and was interested in bringing it into Research Computing. She saw clear synergies between geospatial research and the other centers within the unit. That decision was a game changer. It provided a modest but critical level of central funding that allowed the mission of geospatial support to continue.

At that point, U-Spatial was still small—just two staff members, supplemented by graduate and undergraduate students as projects required. It didn't take much funding to remain sustainable, but having stable institutional support, an acknowledgment of the importance of a geospatial infrastructure, made all the difference.

This experience highlights the importance of Research Computing as a framework for building and sustaining a geospatial support center. Today, Research Computing consists of five different units: the Minnesota Supercomputing Institute; Data Science AI Hub; International Institute of Biosensing; GeoCommunities; and U-Spatial. The Minnesota Supercomputing Institute is the largest and provides high-capacity computing resources and consulting to the university and other educational institutions across Minnesota. The Data Science AI Hub is a rapidly evolving center that's still finding its full mission within the university and beyond, as all colleges and universities work toward integrating artificial intelligence (AI) into their campus culture. Research Computing also includes the International Institute of Biosensing, which focuses on human-environmental interactions. This work spans a range of applications, from everyday technologies such as wearable devices to advanced developments in medical monitoring, environmental sensing, and related fields.

The fourth unit is GeoCommunities, a newer center focused on community-engaged research. GeoCommunities uses GIS tools to create new communities and connections among people and organizations at the local and global levels. It also offers seed grants that faculty and staff can use to support GIS-based strategies related to their research in various parts of the world.

What distinguishes Research Computing and the fifth unit, U-Spatial, from many similar centers at other institutions is that it's fundamentally a support organization. We operate within an academic environment and receive limited central funding, but our mission isn't driven by a stand-alone research agenda. Instead, the role of the Research Computing units is to support and collaborate with researchers and learners across the university. We provide research computing expertise—in high-performance computing, biosensing, or geospatial analysis—to projects across disciplines and domains. Our work frequently extends beyond the university, involving collaborations with government agencies, nonprofit organizations, and partners around the world. Although we help bring funding into the university, we are often not serving as principal investigators. More often, we contribute specialized geospatial expertise as part of larger research efforts, supporting projects as needs arise.

Not having a research agenda is a key factor in the success of U-Spatial. Instead, the success of U-Spatial comes from its mission and vision: to support and collaborate with students, faculty, and staff around the university on GIS, remote sensing, and spatial computing. This is different from the mission of a traditional research center that's driven by discovery. It's also different from the motivations of individual faculty members, who are often focused on tenure, publications, and getting grants.

By having professional staff whose role is to collaborate with other units across campus and provide spatial expertise, you create a different value system, a different reward structure and measure of success. That's another takeaway: Geospatial support is best done with professional staff. And that doesn't mean professional staff alone. U-Spatial today is a group of 16 staff members, with multiple graduate and undergraduate research assistants at any one time. We still have a mission to train students, to give them real experience working on projects in the GIS and geospatial world.

Rapid expansion (2019–2026)

When U-Spatial started, it began modestly, with one staff member and one graduate research assistant in the central core. In the first several years, we expanded by hiring more research assistants (RAs)—students—to provide geospatial expertise to staff across the university. Depending on your institution, that model would work well. By starting small and expanding the center, you can stay flexible and adjust as opportunities occur.

There are limitations, of course. Students have this inconvenient habit of graduating and going on to get amazing jobs, so staffing becomes a revolving door of bringing in new graduate or undergraduate RAs and making sure they have the training, expertise, and resources to succeed. Don't underestimate the cost of working with students. And don't underestimate the reward for the staff, the students, and the university in training the future workforce through real, hands-on projects.

In 2019, U-Spatial merged with the Geospatial Analysis Center (GAC), located on the Duluth campus. The GAC has existed far longer than U-Spatial, and although it supported the students, faculty, and staff on the Duluth campus with their GIS needs, it was primarily a consulting group focused on hazard mitigation planning for local units of government from across the state, which it continues to do to this day. A pitch to merge the two centers was accepted by the vice president for research, and in a short time, the separate units gelled as one. The new U-Spatial had doubled in size to a team of six people, but the multiplier effect of this merger was much larger. The bigger team filled skill gaps and created synergies that allowed U-Spatial to collaborate on more and larger projects. As a result, one to two new staff positions were added each year as the backlog of projects grew.

In the last few years, we have made efforts to include geospatial literacy as part of the university's core competencies, with skills such as reading, writing, and critical thinking. So far, we haven't been successful in making geospatial literacy part of the official core curriculum, but many instructors have incorporated spatial thinking into their courses. Whether through story maps or spatial analysis projects in ArcGIS Online, this approach has been adopted across disciplines, including history, modern languages, design, landscape architecture, remote sensing, and even the business school. Chapters 2 ("Research") and 3 ("Teaching") are full of great examples and case studies.

We have moved from teaching GIS to data science, where geospatial data and analysis are now recognized as a specialized branch of that broader field. We're teaching AI, continuing with machine learning, and putting greater emphasis on scripting with Python and R. And now that geospatial thinking isn't confined to geography, we're meeting needs across departments, from public health to forestry to design.

Research is teaching. It's a simple, maybe even obvious statement from a previous VP for research at Minnesota. Internships and working on real-world projects, what we call experiential learning, is valuable to a student, the faculty, and a future employer. When searching for research funding, we should also frame the effort as capacity building and preparing tomorrow's workforce with students, grad and undergrad—getting real-world experience with projects for the private sector, government, and NGOs.

On the surface, what is seen as U-Spatial today is perhaps a small part of the vision from when it started with just the central core. But the heart of the other three cores has been integrated into the work, and we can confidently and proudly say that the investment in GIS has created the Spatial University.

What services does a geospatial support center provide?

You can read more about geospatial support centers in chapter 7 ("Building a Geospatial Support Center"), but for now let's look at what services are important to offer at your institution. These core services are the same for small to large higher education institutions, although the implementation of the delivery of services will vary greatly and need to adjust to the individual culture of your institution. We believe this portfolio of services is important to sectors outside higher education as well. A large construction company or local government agency would also benefit from offering the same services to its staff.

It's important to note that a single center or department doesn't need to deliver all these services; in fact, some decentralization allows individual departments or units to focus on their core strengths. Having strong coordination among the multiple departments supporting GIS is helpful and prudent.

Help desk

Your students, faculty, and staff are going to have questions about GIS. A help desk can take the form of a physical location where people can drop by or a virtual location through email or messaging apps. Help desk materials should focus on information specific to your institution and point to self-service resources when it makes sense. For example, when someone sends an email to our help desk, an automated reply provides information on how to sign in to ArcGIS Online, which is the most common question we receive.

Search engines and AI are becoming powerful tools for answering many questions, but they cannot replace expert guidance at key times. For instance, a student might be interested in mapping the locations of urban beekeepers to ensure there is enough space between hives to prevent overpopulation. The student, through searching and AI, discovers a new tool, ArcGIS Pro. ArcGIS Pro would certainly work in this instance, but the time needed to learn the software could be a barrier. A quick conversation with the help desk staff allows the student to describe their research objective and very likely learn about ArcGIS Online as a more appropriate tool for analysis with less time needed to learn desktop software. During the conversation, the student also learns about ArcGIS StoryMaps and gets excited about using that app to share results.

Staffing the help desk can be achieved by using staff, students, or both. If your campus offers a GIS major, you probably have a ready workforce of students who can staff the help desk, which also exposes them to many types of GIS questions that will sharpen their expertise.

Training workshops

Once your colleagues—students, faculty, or staff—are aware of GIS, they'll need training to understand the methods and software. Many institutions teach introductory GIS courses, and although it would be amazing if every student could take a semester-long GIS course, many degree programs don't have the capacity to add just one more class.

The amount of online materials for learning GIS can be overwhelming. Online searching and AI chatbots are often a good way to get started. These search options can lead to a wealth of resources, including Esri Community (a global community of Esri users) and Stack Overflow, that many colleagues at higher education institutions freely provide online as course materials and library guides. Massive, open online courses (MOOCs) are another great resource for learning GIS. Esri Academy, the company's comprehensive training center, offers MOOCs on such topics as spatial analysis, GIS for climate action, imagery, and spatial data science.

We don't promise to teach students a software package and Python or R to get a job, but rather the critical thinking (problem solving) skills to learn new tools and techniques and to know when to take advantage of disruptive change like AI. There is no reason to believe that change is going to stop or even slow down.

Students can access many resources online at their own pace, but we have found that students still have a strong desire for instructor-led synchronous workshops. Whether in person or online, having time to interact and ask questions is helpful for many learners and often leads to future collaborations.

That's where workshops can fill the gap. Workshops can introduce aspects of GIS, such as web mapping with ArcGIS Online and storytelling with ArcGIS Story-Maps. Workshops are also highly effective in building specific skills, such as geocoding or field data collection, with tools such as ArcGIS Survey123 and ArcGIS Field Maps. These introductory workshops can be two to three hours long and give learners the foundation to ask questions and learn skills as needed. Spatial is special, and we believe that as people start using GIS, they need to have some understanding of core geographic principles.

Offering workshops takes planning and commitment to keep materials current. Think about the topics that are important to your campus—for example, satellite imagery, high-performance computing, or GeoAI (AI-driven geospatial workflows). Check the resources cited in this book for the many materials you can incorporate into your workshop. As a workshop portfolio has developed, we have seen strong interest from learners in obtaining microcredentials, such as badges, by completing a workshop series. The badge provides evidence that learners have basic skills that can be included on a résumé. At the University of Minnesota, our Geospatial Storytelling badge has been popular.

As your workshop offerings expand, the next level would be to create a boot camp that can provide intensive, focused training in a short time. Great examples of this approach include the Yale Geospatial Accelerator, Harvard Summer Institute, and the Institute for Geospatial Understanding Through an Integrative Discovery Environment's annual I-GUIDE Summer School.

Support for instructors and learners

Spatial thinking can benefit nearly every discipline on campus, as we'll show in the following chapters. Successfully integrating GIS into the classroom requires a commitment to support and collaborate with instructors and students using a three-step process. The first step is exposing both groups to what's possible when thinking spatially; this awareness can happen through workshops, social media posts, word of mouth, and even visiting countless department meetings. As part of the second step, once they're aware of GIS, instructors and students may need help brainstorming how GIS may be used in their own work. For instance, geocoding, the process of assigning latitude and longitude to an address, although a staple of GIS classes, might be viewed as magic to an instructor in pharmacy who's interested in the drug supply chain. The third step is to work with the instructor on obtaining GIS skills, finding data, and acquiring suitable software to complete analyses. These resources are listed in the book as well.

Supporting instructors and learners requires significant effort in staff time. For an instructor to adopt story maps in their course as a replacement for a traditional term paper requires a willingness to try new technologies and become comfortable enough with them to lead students through various exercises or tutorials. Academic tech workers provide this service on many campuses and supply a useful model for extending GIS expertise.

Access to tools, software, and computing infrastructure

The list of GIS software is long and diverse. That's part of the challenge for someone just starting to think spatially—there's no shortage of options for beginning an analysis. One important service of a support center is to act as a concierge and help people find the appropriate tools and software.

Since it was released in 2011, ArcGIS Online has become the go-to tool for getting started in GIS. A geospatial support center should probably provide administration assistance with ArcGIS Online. During the last decade or so, a dramatic shift has occurred from desktop-based GIS software to cloud-based services for most GIS users. These cloud-based services, such as ArcGIS Online, require significantly less time to make a map or perform spatial analysis, not to mention the ease at which maps can be shared through a simple URL. Desktop GIS software is also being replaced by scripting languages such as Python and R, which are driving the power of GIS into data science domains.

Although the default is to use ArcGIS Online as the primary resource for storing spatial data, some situations require data to be stored on alternative platforms, such as ArcGIS Enterprise Server. For example, when working with personal health information, data and analyses need to be kept in a highly protective environment. We recommend collaborating with your campus IT security experts. A geospatial support center can also help with GIS-specific information.

Access to data

Finding and preparing data to use for mapping or analysis can be a time-consuming endeavor. Many data resources are available, including ArcGIS Living Atlas of the World, the Big Ten Academic Alliance Geoportal for institutional data, IPUMS (originally the Integrated Public Use Microdata Series) for population in the United States and other countries, data portals for every US state, and many other resources. In the United States, the National Spatial Data Infrastructure (NSDI) provides standardized themes for sharing data across all levels of government. Other nations have varying levels of data infrastructure, as described in the *Geospatial Knowledge Infrastructure Readiness Index Report 2025* from Geospatial World. The United Nations Committee of Experts on Global Geospatial Information Management (GGIM) created and promoted the Integrated Geospatial Information Framework as a resource for evaluating the availability of and accessibility to spatial data. In addition, the availability of remotely

sensed data from satellites and the Internet of Things may quickly overwhelm someone learning GIS—a raging river of information. One can hope that the promise of AI
will be able to help tame the river of data.

The libraries at most institutions are well suited to providing support for working with spatial data. Remember that although students, faculty, and staff may need
help finding data, they are also data producers through their research and may need
direction on how to properly manage data, as well as ensuring that data is made available, archived, and preserved for future use. Two overarching frameworks provide
guides for our colleagues: FAIR (findable, accessible, interoperable, reusable) provides
guidance on making data machine-readable for reuse, and CARE (collective benefit,
authority to control, responsibility, ethics) provides guidance on data sovereignty and
ethical use.

Community building and outreach

The successful geospatial center is a cheerleader for how and why GIS is important
to students, faculty, and staff at an institution. Many of us have heard a student in
their senior year say, "I wish I would have discovered GIS earlier." Whether the student wants to switch to GIS or just see how they can use it in their work, there should
be a spectrum of opportunities available to learn and apply geospatial fundamentals.

Building a community is time-consuming and requires investment in multiple levels of engagement to be successful. Think about events that may have broad interest
to the entire campus, such as events on GIS Day (held annually on the third Wednesday
of November), which seek to build awareness of thinking spatially. A speaker series
can attract colleagues who are using GIS by presenting novel ways to use these technologies in research. Creating that community is important when building the case to
administrators about the importance of GIS when funding is sought.

Outreach includes the connections we're making with organizations beyond our
campus. As a land grant university, this university is mandated and funded to work
with the residents of the state, but we suggest that all higher education institutions
are members of their community (and beyond) and realize the benefit of collaborating
with a variety of partners.

Although we're in the business of granting degrees, we also need to increase the
adoption of certificates and badges to increase competence in targeted geospatial
knowledge and methods. In partnership with industry, government, and nonprofit sectors, these certificates can drive retraining, the constant need for learning that a geospatially enabled workforce requires.

Consulting

The final service is consulting—providing expertise and services for a fee or some
other compensation. We joke that if someone contacts the help desk and asks for
assistance learning about GIS, we'll work with them for extended hours at no cost. If

someone contacts the help desk and asks whether we can do the GIS work for them, that's consulting and there will need to be compensation. Speaking from experience, defining the threshold of free support versus consulting is something we recommend defining early and clearly. When U-Spatial started, we provided a great number of hours to a project at no cost because of central funding. As the funding model shifted, it was difficult to change expectations, with a free service now being charged for. Having some level of cost recovery for consulting reinforces the value of the service and is key to building a sustainable support center.

This is not to say that free or reduced-cost services have no place. For example, we'll work with faculty on grant proposals to provide GIS analysis and maps at no cost; being able to make a proposal stronger is good for the faculty, as well as the university. We also realize that some graduate students may not have funding to support GIS services, so we'll work with them by combining GIS teaching and consulting services.

Finally, consulting projects are fun for support staff. The opportunity to learn a little about mapping monkeys in a national park in Uganda or the best location for solar installations in Duluth, Minnesota, are just two examples of worthwhile projects that staff can help advance.

Conclusion

In this chapter, we suggest that GIS shouldn't be seen as a niche technical skill but as a foundational way of enhancing teaching, research, and impact across the entire institution. Spatial thinking is as important as reading and math for deeper understanding, connection, and problem-solving, rather than an end in itself. GIS can bridge disciplinary silos and bring people together around a shared understanding of place. Through our experience building U-Spatial, we demonstrate how a service-oriented geospatial support center can sustainably democratize access to GIS by supporting and collaborating with colleagues to deliver teaching, research, outreach, and workforce development. Ultimately, investing in geospatial support delivers a strong return on investment by transforming institutions into spatially fluent campuses that use the power of where to address complex, real-world challenges. The following chapters provide examples from across academia and beyond of what's possible when spatial thinking is widely embraced and integrated.

CHAPTER 2

Research

Introduction

The research community widely uses geospatial tools, with applications in virtually every discipline. Research helps us understand what *is*, *how* it works, and *why* it works as it does, accomplishing this through a systematic process of observation, hypothesis formation, and experimentation to test that hypothesis. This procedure is followed by the analysis and communication of the results to one's peers, whose critical assessment of the work leads to further rounds of research in a never-ending cycle of knowledge creation. GIS adds to this process an understanding of *where*: where phenomena have occurred, are occurring, or may occur in the future. Everything happens somewhere; in many cases, location matters in helping us understand what happens, how it happens, and why.

This chapter considers GIS-related research from several perspectives. It looks at how GIS supports different types of research, the diversity of research methods and approaches, the range of research funding sources and mechanisms, and the variety of settings in which research occurs. Mapping the research landscape, in other words, can help us navigate it, just as geographic maps can help us find our way around the physical landscape. As we see throughout this book, maps serve as metaphorical as well as material aids, and that's especially true in the research world.

Research types

We typically categorize research into three different, but related, types:

- **Basic** research seeks to establish fundamental knowledge about reality without necessarily thinking about its practical application. As an example, the legal deserts map[1] of the United States by the National Center for State Courts illustrates basic research at its best. It shows where the greatest barriers to accessing legal services exist and how these barriers relate to people's physical location, language ability, digital capacity, and broadband access.

- **Applied** research, in contrast, has a practical purpose in mind. Although it can be just as rigorous as basic research, applied research has the goal not only of understanding but also of changing or improving a situation, whether through persuasion or through the development of new technology, methods, or products. Consider the story map on the Sundarbans,[2] the world's largest mangrove forest, in the Bay of Bengal. Based on a thorough understanding of that diverse ecosystem, the research lays out, in practical terms, the threats facing this forest and what we need to do to protect it.

- **Action** research represents a third type, which often involves collaboration between researchers and community members in addressing a problem, working together to solve it, and gathering data afterward to assess the success of these actions. GIS is especially helpful in this regard, because maps have always served to show us where we need to go and how to get there. An example of this advantage is the story map on the interactions of people and elephants in Tanzania, which describes how scientists have sought to understand human-wildlife conflicts. The story map shows how research has revealed ways that communities can reduce such conflicts and how researchers have worked with Tanzanian farmers to take the recommended actions.

Research methods

Researchers typically follow one or more of three methods:

- The **quantitative** method involves gathering data, testing hypotheses, analyzing results, and generalizing about them. This method remains perhaps the most common and most associated with scientific research. GIS aids this work by helping us visualize data and understand where that data came from or where it has the greatest relevance. Tucson's Equity Priority Index exemplifies a quantitative research approach, using data about factors such as income level or housing burden to identify where the city should prioritize its investments.

- The **qualitative** methods in the social sciences, such as anthropology, provide understanding through observing, listening to, and interacting with the people in a place through interviews, focus groups, and surveys. GIS tools such as ArcGIS Survey123 readily support such methods. The story map of conversations with Indigenous people about land rights in South America shows the power of qualitative research to capture the nuances of tribal communities and their relationship to the land they've occupied for thousands of years.

- **Mixed** methods in research combine features of quantitative and qualitative methods as a way to capture both factual data and observed behavior. Story maps offer an ideal vehicle for such research because they accommodate the quantitative visualization of data through GIS maps and the qualitative insights that come through photography and narrative. The story map of the Kuchis, the nomadic people of Afghanistan, illustrates that mixture, with data-based maps of the tribes' migratory routes and the nomads' stories about the struggles they face in meeting their basic needs.

Research designs

Research, like any human activity, involves design, whether it's the design of an experiment, a survey, or a process. Five types of research design recur in the literature as useful ways to proceed:

- **Exploratory** research offers one of the most adaptable designs, well suited to situations in which the problem is either not well understood or clearly defined. Such research often has an improvisational character because the researchers may not even be sure what they're looking for. An example of this is the story map about a geologist's search for Delaware's boundary monuments, whose locations were unknown or long forgotten, turning his research into a monumental scavenger hunt.
- **Descriptive** research often arises out of the exploratory phase, where researchers describe a situation or phenomenon as accurately and systematically as possible, with little or no commentary. GIS offers an excellent platform for such descriptive work by showing data visually, without the need to interpret what it means, evident in the story map of Prague, showing that city's diverse urban patterns.
- **Correlational** research takes that descriptive work into a more statistical direction, focusing not only on typical patterns but also on the statistical likelihood of a connection between two or more phenomena. Geospatial analytics can help reveal those connections, as we can learn from the story map about the removal of dams along the Charles River in greater Boston and the likely impact this removal would have on the water quality of the river and its aquatic life. As the study shows, fewer dams correlate to a healthier ecosystem.
- **Causal** research seeks to add certainty to an experiment or exploration by changing one or more independent variables and measuring their impact on a dependent variable, often in a controlled setting. GIS can help show causal relationships that occur outside a laboratory. The story map of climate adaption in Nelson, New Zealand, shows the impact that various degrees of

sea level rise would have on low-lying parts of the city. The cause and effect of rising seas and greater coastal flooding may not be welcome information, but it's indisputable and can help communities prepare.

- **Causal-comparative** research looks at cause-and-effect relationships after the fact, making comparisons among phenomena that already exist or that might exist in the future. GIS layers can enable this type of design. An example is the story map comparing different transportation scenarios in greater New York City and evaluating the impact on the power grid of electrifying the various transportation options for commuters.

Research funding sources

The amount of research funding available varies considerably among countries, but the types of funding typically do not.

- **National governments** often provide a sizable percentage of the money available for research, particularly for research that has potential public benefit and basic research that the private sector may not fund. Government funding may go to public agencies or universities and research institutes, as in the case of the story map of the Kaibab National Forest, in which the University of Arizona Libraries' Data Cooperative digitized, analyzed, and compared the hand-drawn maps of the US Forest Service to GIS maps.
- **Private-sector** research tends to stay within the companies supporting it for competitive reasons, even when some of this research might be done in partnership with academic or independent labs. In the case of S&P Energy's story map, research into the movement of ships makes the case, through maps, for the strategic role of the Port of Fujairah, which has become one of the world's largest bunkering ports because of its location near the Strait of Hormuz.
- **Foundations and other nonprofit organizations** support research into topics related to their missions, and GIS can facilitate that in multiple ways, from showing where the greatest needs exist to revealing where the organization's mission may have the greatest effect. The story map by the United Nations about the impact of COVID-19 on education, hunger, and social protections in places such as South Sudan shows how a pandemic affects not only human health but also so many of the other factors that lead to a healthy life and a healthy society.
- **Universities and institutes** also support research internally, not only through their funding of research faculty and student assistants but also through seed grants and endowment funds. An example of this is the story map by the Harvard Chan–NIEHS Center for Environmental Health, which examines

the history of racial segregation in Boston and the impact that the spatial
separation of races continues to have on the health of neighborhoods to
this day.

- **Individual donors** represent a final and significant source of research funding,
  whether from wealthy philanthropists or online crowdfunding platforms
  interested in supporting often highly focused or early-stage research
  projects. The Open Space Institute is an example of a donor-supported
  nonprofit, and its story map of its efforts to protect the Shawangunk Ridge in
  New York state shows how research can inspire philanthropy and individual
  contributions to its cause.

Research funding mechanisms

Research funding comes in various forms, each of which has a place in the research
community.

- **Direct grants** in support of research projects that have a specific purpose,
  time line, budget, and deliverable are the most common. National Geographic,
  for example, has sponsored research since 1890 in support of major
  environmental and exploratory efforts, such as the Okavango Wilderness
  Project, exploring and protecting the Okavango River Basin in Angola, Namibia,
  and Botswana.
- **Fellowships or scholarships** to researchers and graduate or postdoctoral
  students typically fund an individual's research or a person's training and
  career development related to their research interests. The story map of
  a National Geographic Explorer's efforts to help save the Amazon River's
  dolphins from extinction shows how one person can make a real difference.
- **Contracts** with government agencies or private companies typically fund
  research for a needed service or a project with well-defined milestones and
  deliverables. The US Congress established the US Agency for International
  Development (USAID) to contract work essential to the health and safety
  of people around the world. As its story map shows, it has helped prepare
  communities for hurricane season in the Caribbean, including those in the
  United States.
- **Cooperative agreements** give funders more involvement in the research
  or enable diverse groups to work more collaboratively. An example of a
  cooperative agreement is the story map of Canada's Confederated Salish and
  Koortenai Tribes, which have worked together to revive Indigenous ways of
  living and to prepare a strategic plan to fight climate change and its projected
  effects on their land.

- **Seed grants**, which often fund the initial stages of a research project or support the search for additional external funding, typically come from within an organization or agency. Such grants can produce outstanding work from motivated student researchers, as in the story map by a first-year graduate student documenting Mexican farmworkers' experience since the early 20th century, as they have moved back and forth across the US–Mexican border.

Research locations

Research occurs in a variety of locations as well, depending on its nature, purpose, and sources of funding.

- **Universities** remain the dominant location for research in most parts of the world because of the diversity of disciplines they house and because most have the infrastructure needed to support research. Much of that research happens in laboratories, although sometimes that infrastructure includes field research, as in the story map that analyzes the size of farms and the diversity of nutrients they produce globally.
- **Academic libraries and archives** offer another, critically important location for research, especially in the arts, humanities, and social sciences, where databases, books, historical documents, and maps provide the basis for much of the work. A good example is the story map on 19th-century silver mines in Virginia City, Nevada, with historical maps from Stanford University Libraries as the basis for creating 3D subsurface GIS maps of the mines.
- **Research institutes** also provide a place for specific types of investigations, often related to but frequently separate from academic institutions. The Dena Kayeh Institute in Canada's British Columbia represents one example: an institute dedicated to preserving the Kaska Dena language, oral traditions, and traditional knowledge, well-illustrated in the story map of that culture's history and traditions.
- **Corporate research and development departments** are the locations for much of the applied research that occurs, often with an eye to the creation of new products and better processes with greater profit in mind. Consider Esri, whose research and development in several centers around the globe has transformed geospatial tools.
- **Hospitals and clinics**, especially those connected to research institutions, house much of the clinical research related to medical procedures, drug trials, and public health studies. An example of the excellent work that can arise from public health research is an article by a group of Canadian researchers mapping the locations in Ontario where the risks of traumatic brain injury[3] are the highest and how the causes differ, depending on place.

- **National parks, ecological preserves, and wilderness areas** are also research sites for disciplines such as environmental science, biology, and climate change studies. Such places can provide exceptional opportunities for field research, as we can see in the story map by the National Park Service of the diverse ecosystems in San Juan Islands, Washington state.

- **Community and public spaces** offer other types of field research sites for disciplines such as social science, urban planning, and ethnography. Such locations have the added benefit of accommodating citizen science, in which the public becomes a part of the research team, as happened in the citizen science project in Charlotte, North Carolina, mapping air pollution near the city's highways.

- Finally, much research occurs virtually, on **online platforms**, enabling people physically distant from one another to collaborate on projects as never before. GIS enables this in various ways, including, for example, the use of satellite imagery and OpenStreetMap information to understand food insecurity in northern Uganda or the use of surveys and volunteered geographic information to locate the households in Chad with children in need of measles vaccination.

In all this work, where things occur matters as much as how, when, and why they occur, making geospatial tools an essential part of most research, as the following examples show.

Research type examples

Human-elephant conflicts

DISCIPLINE Wildlife   **INDUSTRY** Conservation

ORGANIZATION African People & Wildlife   **LOCATION** Tanzania

Conflicts between farmers and wildlife venturing into their fields have resulted in the killing of animals and destruction of property. To avoid that outcome and protect wildlife such as the elephant herds in northern Tanzania, researchers with the African People & Wildlife and the Ngorongoro Conservation Area Authority (NCAA) have mapped instances where elephant herds have invaded farmland and have equipped farmers with "coexistence kits," and the farmers have learned to use nonviolent strategies—bright or flashing lights, air horns, or the smoke from burning dung and pepper—to ward off the elephants. By understanding the behavior of elephants and the concerns of farmers, the research team has developed several low-cost and effective ways of protecting both.

Legal deserts

DISCIPLINE Law

INDUSTRY Local Government

ORGANIZATION National Center for State Courts

LOCATION Minnesota, USA

Deserts can take many forms, wherever there's a lack of something essential to life, whether it's water in a literal desert or access to legal aid, as in this map by the National Center for State Courts. That center has mapped 20 US states in terms of the number and location of attorneys, the locations of courthouses, the estimated distances people must travel to get to them, and the ratio of attorneys to the population in each state. The project includes data layers that map population density, English language proficiency, poverty, and internet access and broadband availability. This mapping of legal deserts in the United States shows that although every American can have their day in court, getting to and from court and finding someone to represent them in court presents a real challenge for many people.

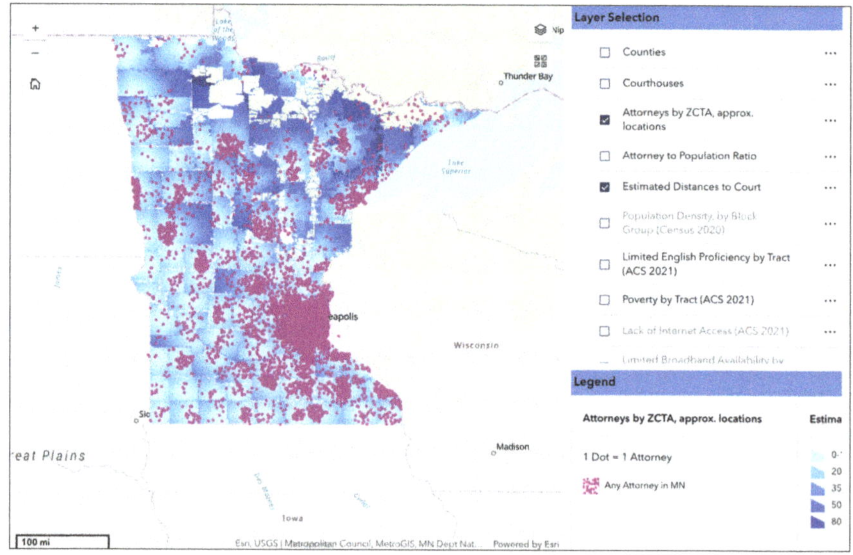

Although Americans all have the right to due process, many people live far from courthouses, with few attorneys close by. *Courtesy of National Center for State Courts.*

The Sundarbans

DISCIPLINE Forestry   **INDUSTRY** Natural Resources

ORGANIZATION Student   **LOCATION** India

Mangrove forests offer multiple environmental benefits, ranging from the provision of habitat to the sequestration of atmospheric carbon to the protection of coastal areas from increasingly intense storms and rising sea levels. The Sundarbans on the Bay of Bengal, southeast of Kolkata, India, covers 10,000 square kilometers and remains the largest mangrove forest in the world. In recent decades, it has been damaged by several severe cyclones and increased soil salinity, leading to the death of many mangrove trees. This story map by two researchers documents the ecological and economic value of the Sundarbans and the impact that climate change has begun to have on it, calling for a co-management strategy of the forest by both India and Bangladesh.

The Sundarbans on the Bay of Bengal is the largest mangrove forest in the world, and it faces several threats to its future. *Courtesy of Shwarnali Bhattacharjee, Hasan Ahmed.*

Research method examples

Territorial justice

DISCIPLINE Public Policy

ORGANIZATION Amazon Conservation Team

INDUSTRY Conservation

LOCATION Colombia

The Indigenous populations in the Amazon and across the Andes in the northern part of South America have a rich cultural tradition tied to the land. This story map describes how the various tribal communities in the region have coped with incursions into their territories by mining and logging interests and what they can anticipate, based on the past 20 years, with the Amazon bioregion having lost about 13 percent of its forest cover. That loss of forest can destabilize the Amazon's natural water cycles and permanently damage the rain forest on whose health the whole world depends. The story map also tells compelling stories of how Indigenous people have restored damaged landscapes and stewarded the forests, giving us hope that as Indigenous people pursue more rights-based conservation models across the Amazon basin, we'll all benefit from the health of that globally significant ecosystem.

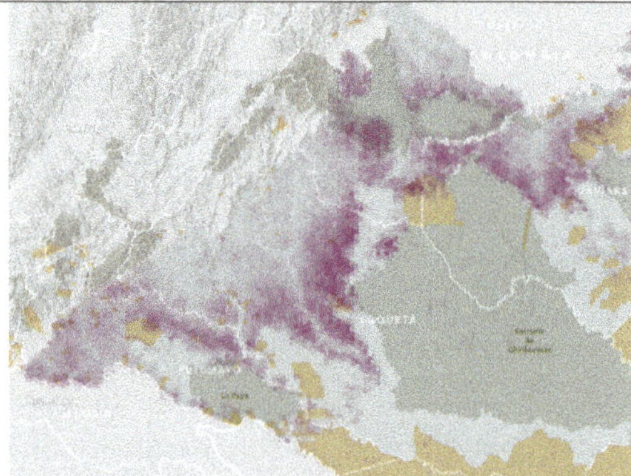

The Indigenous people of the Amazon have retained many of their traditional cultural practices, some of which are threatened by deforestation. *Courtesy of Amazon Conservation Team.*

Nomadic tribes

DISCIPLINE History

ORGANIZATION Alcis

INDUSTRY Sustainable Development

LOCATION Afghanistan

The UN's Sustainable Development Goals provide a framework that can help countries and communities address the greatest needs of people. However, when the people most in need lead a nomadic life, as is the case with the Kuchi tribe in Afghanistan, serving those needs can be challenging. This story map uses aerial imagery to track the path that the Kuchis follow every year as they migrate from the central highlands in search of winter pasture for their sheep and goats, following a major river system in Herat Province, Afghanistan. The researchers lay out the challenges that the Kuchis face, including a lack of food, health care, and education. The researchers hope that, by knowing when and where the Kuchis migrate, organizations can address their needs more effectively. It's a moving target but one worth pursuing.

Equity priorities

DISCIPLINE Public Health

ORGANIZATION City of Tucson Equity Priority Index

INDUSTRY Local Government

LOCATION Tucson, Arizona, USA

The physical fabric of a city may not look much different from one neighborhood to another, but when you factor in the social vulnerability of people, places can differ widely in their resilience. Tucson, Arizona, has developed an equity index that maps the social vulnerability of its citizens, measuring people's ability to "anticipate, cope with, resist, and recover from external stressors." A social vulnerability index can "include an area's economic status, health care, and the prevalence of chronic disease, educational attainment, housing affordability and homeownership rates, and social factors like age, language, barriers, and population density." Such an index enables the city to identify the neighborhoods where its services are most needed. This publicly available story map allows Tucson's residents to filter the data according to their neighborhood or by various factors, making access to this equity index truly equitable.

Research design examples

Urban diversity

DISCIPLINE Urban Planning   **INDUSTRY** Local Government

ORGANIZATION Prague Institute for   **LOCATION** Czechia
Planning and Development

Cities often reflect the political cultures that oversee their development, and we can read those cultural differences in the form of the city, especially if, as in the case of Prague, capital of Czechia, the city has gone through dramatically different types of governments and economies. This story map by researchers in the city looks at urban patterns from the pedestrian-oriented medieval Old Town and the tram-dependent 19th-century neighborhoods to the suburban garden city of the early 20th century and the repetitive housing blocks of the Soviet era. The authors also look at the different ways in which people view the city as pedestrians, bicyclists, and drivers and how that affects perceptions of the diversity and scale of the city. This map shows how a city such as Prague can act as a mirror to who people once were, how they now live, and what they want to be in the future.

Climate change preparation

DISCIPLINE Urban Planning   **INDUSTRY** Local Government

ORGANIZATION Nelson City Council   **LOCATION** Nelson, New Zealand

Many cities occupy coastal locations for good reason, but with climate change likely to produce record levels of flooding as sea levels rise, those coastal locations have become a liability. The city of Nelson, New Zealand, has begun to prepare its population for a future in which a substantial portion of the city will be below water, depicted in the story map of the areas of the city most affected by a two-meter rise in sea level. City councils in many coastal locations have not been as forthright as Nelson's about the situation they face, perhaps not wanting to alarm their residents or depress land values. But making the impacts of climate change visible and local may be one of the best ways to begin finding solutions, as Nelson has done by looking to the community for ideas about specific adaptation options for various locations. In this case, preparation is the best policy.

Boundary markers

DISCIPLINE Civil Engineering

ORGANIZATION Delaware Geological Survey

INDUSTRY Local Government

LOCATION Odessa, Delaware, USA

We assume we know where boundaries lie, but the actual locations of boundary markers can show the lie behind that assumption. This story map recounts the research by one of its authors, a geologist, locating the 179 stone boundary markers for the state of Delaware originally set in the 1760s, part of which marked the Mason–Dixon Line, which became the symbolic dividing line between the free and the slave states during the US Civil War. Locating the markers proved challenging in places where later development or changes in the land had occurred, and the research showed discrepancies between the boundaries shown by official maps and the markers. The work reminds us of how hard it is to draw lines on ever-changing landscapes and how much our ideas about the world can vary from the reality of it.

The differences between boundary lines and boundary markers highlight the gaps between policy and reality. *Courtesy of Lillian Wang, William Schenck, and the Delaware Geological Survey.*

Dam removal

DISCIPLINE Environmental Science

INDUSTRY Water

ORGANIZATION Charles River Watershed Association

LOCATION Waltham, Massachusetts, USA

The damming of rivers has been a practice that communities have adopted to generate hydropower, control flooding, and—in the case of the Charles River in Massachusetts—oppress the local Indigenous population dependent on the river for food. This story map documents the history of dams along the river and their devastating effect on migratory fish, which lost two-thirds of their former habitat once the dams were built. The Charles River Watershed Association calls on the state and the communities along the river to remove the dams, joining 60 other dams that have been removed across the state. With dams no longer used to generate power and causing so much environmental damage and flooding induced by climate change, their removal seems like just a matter of time. When it comes to dams, the dam of political opposition to removal is finally breaking.

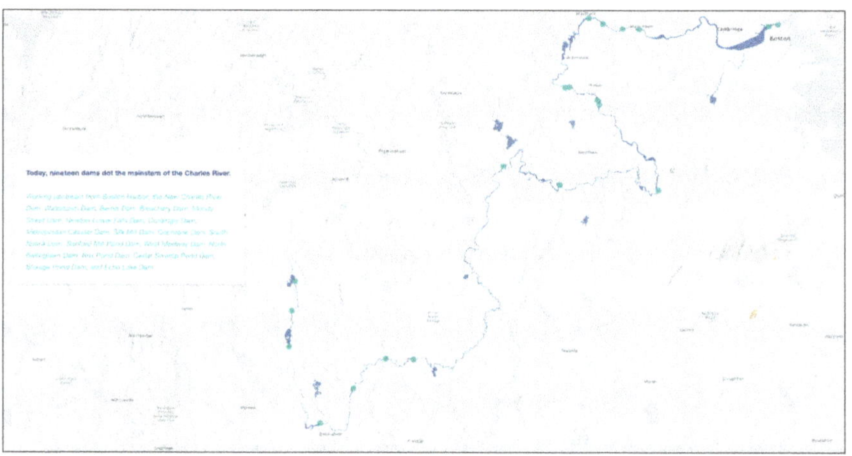

Dams damage the ecology of rivers, and their removal has become one of the most effective ways to restore regional ecosystems. *Courtesy of the Charles River Watershed Association.*

Electric commuting

DISCIPLINE Public Policy

INDUSTRY Transportation

ORGANIZATION New York University
Center for Urban Science
+ Progress

LOCATION New York, New York, USA

Transportation produces more than one-quarter of the US greenhouse gas emissions, so electrifying commuter transportation offers an opportunity to greatly reduce that pollution. This story map by researchers at New York University considers what a fully electric transportation system would look like for New York City's three million daily commuters. The researchers created a number of commuting scenarios and apportioned the transportation modes for each. They then looked at the ability of the power system to handle a fully electric transportation system. The research team also examined a few changes, such as more bike-friendly bridges, that would dramatically affect the use of modes other than cars. A smartphone-friendly dashboard also gives commuters a better sense of the impact of their transportation decisions and the public policies aimed at reducing transportation pollution. It's electrifying work.

Research funding source examples

Historical maps

DISCIPLINE History

INDUSTRY Education

ORGANIZATION University of Arizona
Libraries Data Cooperative

LOCATION Arizona, USA

Historical maps appeal to us in their combination of graphic quality and aesthetic sensibility, but they're also a record of how people in the past came to understand a landscape long occupied by Indigenous communities. This story map of the maps and history of the Kaibab National Forest in northern Arizona recounts how native people saw it as their spiritual home, how the US government saw it as its responsibility to protect, and how forestry and forest fires have come to dominate the landscape in modern times. The digitizing of the hand-drawn Forest Service maps has also allowed us to compare earlier records of borders, fires, and timber to modern satellite imagery and, in the process, to protect the historical record. The time has come to conserve the maps of America's earliest conservationists.

Racial segregation

DISCIPLINE History

INDUSTRY Education

ORGANIZATION The Harvard Chan NIEHS Center for Environmental Health

LOCATION Boston, Massachusetts, USA

Not only was racial segregation allowed by the US government throughout much of the 20th century, but the government mandated it by discouraging loans in poorer neighborhoods by federally insured banks. Although redlining was outlawed in 1968, the long-term impact of those discriminatory policies remains to this day in cities such as Boston, shown in this map by researchers at Harvard's School of Public Health. The red areas show where banks avoided making loans, and the yellow areas show where they used extreme caution—both areas where people of color and poorer working families lived. The irony is that, although a sizable majority of the US population lives in metropolitan areas, much of the national policy and political discourse is hostile to cities and the people who live in them. The tragedy of redlining was both a racial and an economic one, as banks refused to invest in the very people who are essential to a healthy economy.

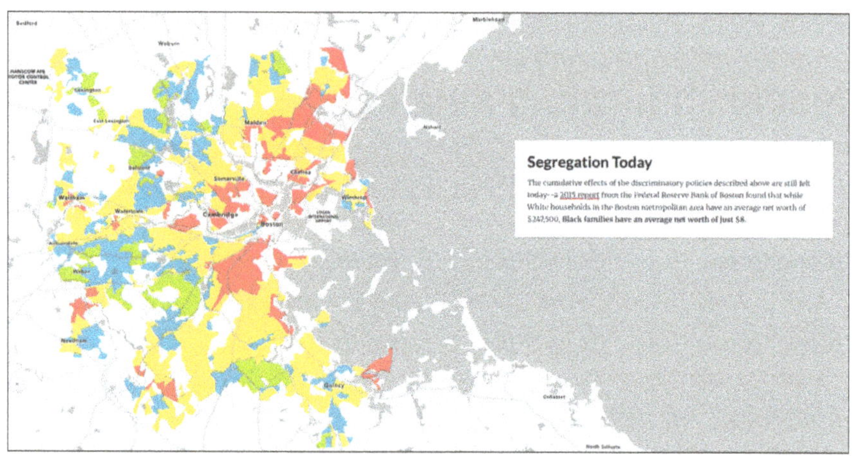

Redlining, which allowed banks in the United States from the 1930s to 1968 to discriminate against the poor, has left a lasting legacy on many neighborhoods today. *Courtesy of Harvard Chan, NIEHS Center for Environmental Health.*

Port positioning

DISCIPLINE Economics   **INDUSTRY** Business

ORGANIZATION S&P Global   **LOCATION** UAE

Maps help us make a point about a place as much as they point us to the places we want to go, and that dual purpose is evident in this story map about the Port of Fujairah in the United Arab Emirates. Produced by a global financial company, the map shows how the port has grown dramatically over the last four decades and how its location on the Arabian Peninsula has made it a key location for ships that may not be able to pass through the Strait of Hormuz. The story map also shows how oil pipelines have been routed to the port to enable ships to dock, making it one of the largest bunkering ports in the world. The great irony of maps is that they serve purposes at odds with each other; this map illustrates an example, showing how the efficient shipping of oil works against the reduction of fossil fuel and the mitigation of climate change explored by other examples in this chapter.

The impact of COVID-19

DISCIPLINE Public Health   **INDUSTRY** Sustainable Development

ORGANIZATION UNHCR, The UN Refugee Agency   **LOCATION** South Sudan

The impact of the COVID-19 pandemic, like previous global pandemics, will affect people around the world for decades to come. At the same time, the immediate impacts continue to be felt, especially in countries already dealing with civil conflicts, such as South Sudan. This story map by the UN's Refugee Agency tracks a number of factors affected by the COVID-19 pandemic, including food insecurity in South Sudan, where much of the nation is experiencing a food crisis, where large numbers of people don't have enough to eat, and where food supplies are unpredictable or unavailable. During the pandemic, those numbers were extraordinary: 11.4 million people didn't have enough food, especially in and around the locations of camps housing people displaced by the violence in their country. If maps can be called sobering, this is certainly one of them.

Open space

DISCIPLINE Public Health   **INDUSTRY** AEC

ORGANIZATION Open Space Institute   **LOCATION** Kingston, New York, USA

Those who support the preservation of open space have to constantly defend it against those who see the land as something to profit from through its development. An example is the work of the Open Space Institute, which has worked successfully since the early 1970s to protect the Shawangunk Ridge north of New York City from intrusive hotel and residential developments and to allow its continued use as a publicly accessible open space for hikers and bicyclists and a safe place for species to thrive. Over four decades, this nonprofit, in partnership with the State of New York, has used its research to show the value of preserving what is now nearly 33,000 acres of land, a total more than twice the size of Manhattan.

1. In the early 1970s, the **Bashakill**, the largest freshwater wetlands in southern New York, lay unprotected and vulnerable to development.

OSI and the New York State Department of Environmental Protection conserved the 4,000-acre Bashakill, a state-designated Wildlife Management Area and a mecca for birders, kayakers, and hikers.

The Bashakill (Greg Miller)

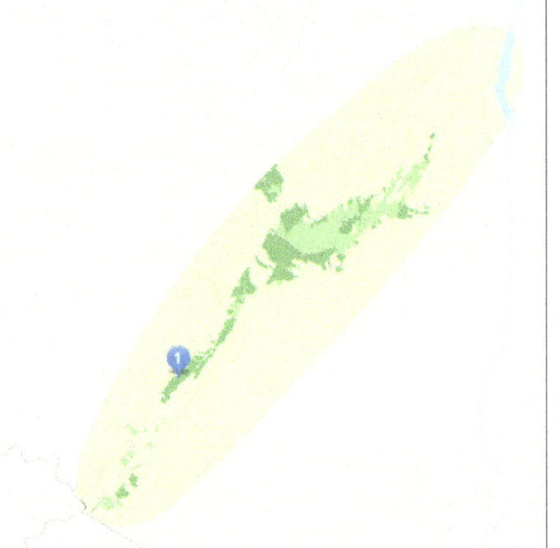

The preservation of open space remains a challenge, especially in thriving metropolitan regions like New York. *Copyright © Open Space Institute. Courtesy of Kelly Proctor and Kae Yamane.*

Research funding mechanism examples

River dolphins

DISCIPLINE Wildlife

INDUSTRY Conservation

ORGANIZATION Omacha Foundation

LOCATION Amazon River basin

Research often reveals the mysteries of the world around us, as these researchers have done by tracking the Amazon River's threatened dolphin population. They used satellite tags to discover where the dolphins feed, mate, and raise their young in Colombia's Lakes of Tarapoto wetland and how widely the dolphins travel through the Amazon River basin. This work is particularly important considering the global threats to river dolphins from habitat loss, dam barriers, water pollution, and loss of prey through fishing—causes that have led to the extinction of some river dolphin species. This story map documents the work of researchers supported by the National Geographic Society and the Omacha Foundation, and it does so in a way that can help the people of the region recognize the importance of dolphins to the health of the river and of their own communities.

Indigenous environmentalism

DISCIPLINE History

INDUSTRY Local Government

ORGANIZATION Confederated Salish and Kootenai Tribes

LOCATION Flathead Reservation, Montana, USA

Although land acknowledgments recognizing Indigenous land that others now occupy have become common in North America, learning from Indigenous environmental practices has been less common. This story map by the Salish and Kootenai Tribes in the Northwest United States explores the traditional knowledge of the Indigenous people of forestry, water, fish, and wildlife. The story map recounts the practices of their ancestors in protecting those valuable resources, shows what tribal communities are doing today to continue those conservation practices, and underscores the threats to the natural environment that result from climate change. What also emerges from this storytelling is the extent to which tribal territories reflect the ecosystems they depend on, a system of occupation of the land that makes more sense than the boundaries that divide us.

River ecosystems

DISCIPLINE Wildlife Biology   **INDUSTRY** Conservation

ORGANIZATION Blue Water GIS   **LOCATION** Botswana

How we communicate can be as important as what we communicate, and that's especially true of communicating conservation efforts. In this story map by Earth-Views, National Geographic's Okavango Wilderness Project and The Wild Bird Trust use GIS, aerial imagery, photography, and video to capture what makes the Okavango such an important world river, providing wilderness habitat in the middle of an otherwise arid Kalahari Desert. The project maps the expeditions that the research community has taken to understand the diverse ecology and hydrology of the place and explores the research questions that remain unanswered. The story map also highlights the negative impacts of poaching, resource extraction, and deforestation in the river's ecosystems and the positive effect of teaching the local population about the economic value of protecting the river.

Immersive story maps can be one of the most effective ways of protecting the planet's most important waterways. *Courtesy of EarthViews, Wild Bird Trust, National Geographic, and Blue Water GIS.*

Migrating farmworkers

DISCIPLINE History

ORGANIZATION Latinos in Heritage Conservation

INDUSTRY Education

LOCATION Texas, USA

Migrant farm labor between Mexico and the United States expanded during World War I. Throughout the 20th century, a series of *bracero* programs provided migrant workers with reliable access to the services they needed as they moved north to perform essential farmwork. This story map documents the various phases of this history and the health screenings that occurred to protect both workers and farmers, as well as the discrimination that migrant workers often faced. Stories of farmworkers' experiences in the United States and how they responded to opportunities and supported one another as they worked make this story map a compelling read, ending with a call to action announcing a grant program to fund similar Latino/Latina heritage projects.

Mexican farmworkers have a long and integral history of moving back and forth over the US–Mexican border. *Courtesy of Latinos in Heritage Conservation, Mel Escobar.*

Hurricane preparation

DISCIPLINE Public Health

ORGANIZATION USAID, Bureau of Humanitarian Assistance

INDUSTRY Science

LOCATION Caribbean

USAID has helped protect US citizens and interests, evident in this story map about its work to prepare the citizens of the Caribbean region for the hurricanes that plague that part of the world. The story map shows how hurricanes have increased in intensity since 1945 and how USAID has helped people build more hurricane-resistant shelter, access emergency aid, and learn how to minimize the disruptions that hurricanes cause. This information is as valuable to US residents living in the path of hurricanes as it is to residents around the world and demonstrates a concern for global health and environmental safety.

Research location examples

Nutrient diversity

DISCIPLINE Agriculture

ORGANIZATION Institute on the Environment

INDUSTRY Natural Resources

LOCATION Global

The size of farms affects the diversity of crops grown. This story map illustrates the range of farm sizes around the world, from very small farms in much of the Global South to very large farms in much of the Global North, with medium-sized farms distributed among both zones. These differences coincide with the diversity of nutrients produced, with smaller farms producing a greater range of crops and larger farms producing food more efficiently. At the same time, small farms produce more than 75 percent of most foods and provide micronutrients for the world's poorest and most vulnerable populations, whereas large farms generate most of the world's supply of oil crops and about half of the sugar crops. Globally, these modes of farming should balance each other, although that depends on effective supply chains and global trade, both of which have been threatened in recent years.

Mine maps

DISCIPLINE History

ORGANIZATION Stanford University Libraries

INDUSTRY Transportation

LOCATION Virgina City, Nevada, USA

Historical maps are typically two dimensional, which makes it difficult to understand three-dimensional landscapes at and below the surface of the earth. The Comstock Lode, a significant seam of silver ore discovered in Virginia City, Nevada, in 1859, was widely mapped at the time. The Stanford University Libraries created this story map to explore historical maps and use GIS to understand the Comstock Lode in a new way. The older maps combined plans and sections and used color to indicate the depth of each mine, but GIS allows for 3D views of the claims that were filed, allowing us to see the relative depth underground of the various mines. It's a fascinating story that leads to some mind-blowing mine maps.

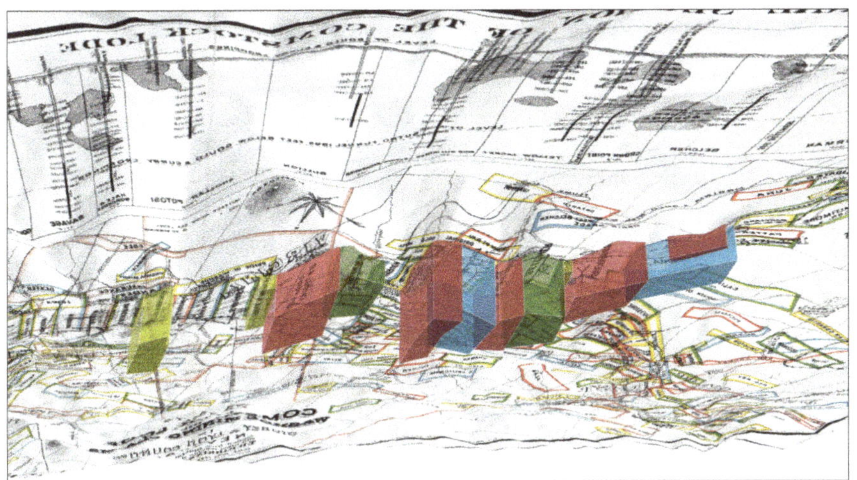

GIS allows us to map subsurface property in visually compelling ways, as in this 3D map of the Comstock Lode. *Courtesy of David Medeiros, Stanford Geospatial Center.*

Brain trauma

DISCIPLINE Public Health **INDUSTRY** Health and Human Services

ORGANIZATION Academic **LOCATION** Ontario, Canada

The occurrence of traumatic brain injuries can coincide with high population areas, but as this research shows, these injuries can also occur in less populated areas, where certain contextual issues, such as the lack of sidewalks or bike lanes, can expose people to injury in more remote locations. The geographic specificity of injuries makes mapping them especially revealing, because GIS can highlight patterns in the data that otherwise may be missed. This map of Ontario, Canada, shows that, although traumatic brain injuries occur more often in metropolitan areas, they can differ in type, with falls happening more frequently in central cities and automobile collisions more often in suburbs. Mapping injuries by a patient's residential address also makes the delivery of long-term care easier.

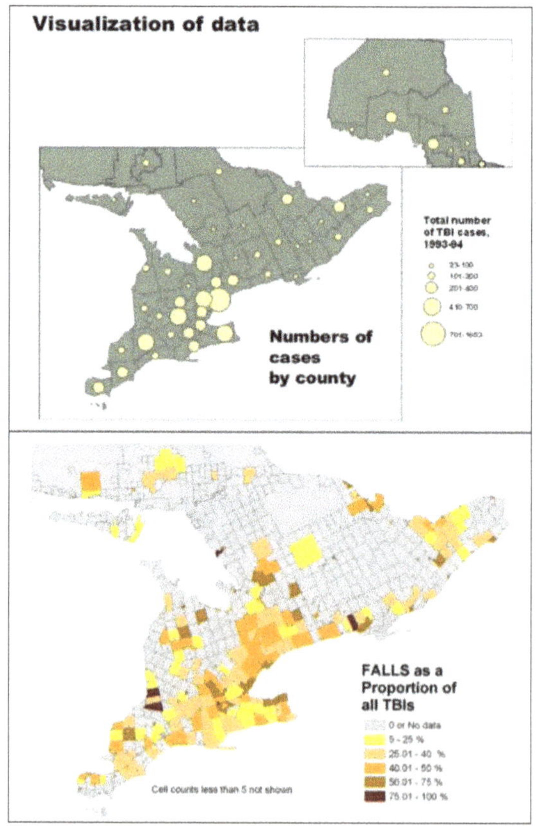

In Ontario, Canada, the incidence of traumatic brain injuries is higher not only in more populated areas but also in pockets where hazardous conditions may exist. *Courtesy of Angela Colantonio, Byron Moldofsky, Michael Escobar, Lee Vernich, Mary Chipman, Barry McLellan.*

Indigenous knowledge

DISCIPLINE History

INDUSTRY Local Government

ORGANIZATION Dena Kayeh Institute

LOCATION British Columbia, Canada

We tend to see Indigenous knowledge as separate from and counter to scientific knowledge, but the Kaska Dena people have shown how both ways of thinking remain relevant. This story map describes the lands called *Dena Kēyeh*—the "people's country" of the Kaska Dena, an unceded nation of people who have lived in their ancestral lands for the last 7,500 years. The *Dene K'éh* is a form of knowledge, known as "the people's way," expressed orally and used to guide appropriate and lasting cultural behavior. The Kaska principles for land and resource management include the recognition of Kaska Dena rights, title, and decision-making powers in their ancestral territory, perpetuated by transmitting their traditional ecological insights, based on all sources of knowledge (scientific, local, and traditional knowledge). It's a mode of thinking that benefits all types of communities.

Research centers

DISCIPLINE Computer Science

INDUSTRY Geospatial

ORGANIZATION Esri

LOCATION Global

Research in the private sector often remains proprietary and only becomes apparent through the products or services that a company makes available to its customers or the public. Take, for example, Esri, which continually invests in research at its headquarters in Redlands, California, as well as in research centers around the world. Each of these centers has a unique research role, ranging from game engine development to 3D reality mapping to AI for geospatial analysis to 3D cartography to field data collection tools. This research, as well as the sense of community it fosters with its customers, has helped Esri remain the largest GIS company in the world.

Air quality

DISCIPLINE Public Health
ORGANIZATION Clean Air Carolina

INDUSTRY Health and Human Services
LOCATION Charlotte, North Carolina, USA

All too often, cities have routed rail and road infrastructure in poorer neighborhoods, subjecting those who live and work there to higher levels of air pollution. In this story map of Charlotte, North Carolina, citizen science researchers and environmental justice advocates mapped the impact of transportation-related air pollution in Charlotte's historic West End, where many African Americans have historically lived. To prove that residents breathe significantly higher levels of air pollution than in other, more affluent parts of the city, Clean Air Carolina collected real-time data with pollution monitors mounted in homes, schools, and public buildings. The research led the city to install a monitoring station and raised awareness of the need for a transportation system that generated less pollution.

Mapping hunger

DISCIPLINE Public Health
ORGANIZATION Seeds for Development

INDUSTRY Sustainable Development
LOCATION Uganda

Food insecurity plagues many people in countries such as Uganda, especially in districts such as Lamogi, where more than one-third of the population eats less than one meal a day. This story map tells how researchers connected to a small nonprofit, Seeds of Development, and worked with vulnerable communities to reach the most food-insecure families by equipping people in each village with a smartphone and ArcGIS Survey123 to locate affected households. The researchers also mapped the location of schools to serve as food distribution hubs and the roads and paths within three kilometers of these schools to estimate the number of households within these catchment areas. The maps helped locate the most remote houses in need of special attention and determine where, in the future, new schools might be located that could also serve as food hubs.

Nature's diversity

DISCIPLINE Recreation

INDUSTRY Natural Resources

ORGANIZATION San Juan Island National Historical Park

LOCATION San Juan Islands, Washington, USA

National parks, one of the great assets of the United States, are not physically accessible to many people without the means to visit them, which makes story maps such as this one on the San Juan Islands so valuable. The story map covers the diverse ecosystems of the islands—forests, prairie, coastal areas, and sand dunes—and accompanies each with a 2D image and 3D model of the relevant eco-systems in both the American and English camps. In addition to information about the species that occupy each of these ecosystems, the story map contains vivid hand-drawn perspectives, photography, and video of the islands, accompanied by a digital twin. For those who never make it to these remote islands in Puget Sound, this map offers the next best thing: an immersive, digital experience based on the best research available.

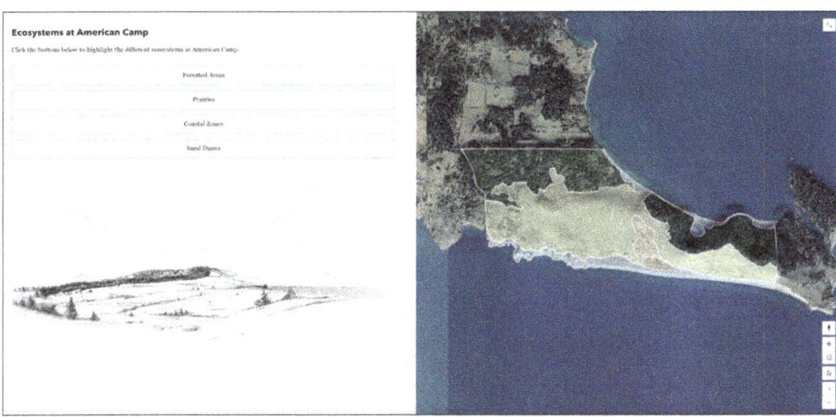

2D and 3D maps can help people understand the qualities of a place and the other species that inhabit it. *Courtesy of San Juan Island National Historical Park, Micah Stailey, Zak Wood, Sara Dolan, Claire Crawbuck.*

Mapping vaccinations

DISCIPLINE Public Health

INDUSTRY Sustainable Development

ORGANIZATION Médecins Sans Frontières

LOCATION Chad

One of the goals of the World Health Organization is to vaccinate 90 to 95 percent of the population to prevent new measles outbreaks. The plan is to make maps a key part of reaching that goal. Médecins Sans Frontières (Doctors Without Borders) uses OpenStreetMap to locate people who need vaccinations, but in many places, the researchers lacked the maps they needed, so they started the Missing Maps Initiative. As part of the initiative, volunteers create maps where there are none, and this story map shows how this effort worked for the Am Timan region in Chad. In about four weeks, 221 volunteers mapped 34,120 buildings in a detailed map of Am Timan and its surroundings, based on hand-drawn maps supplied by the local population. These maps not only helped people get vaccinated against the measles virus but also helped create a sense of community among participants, vaccinating them against the ills of social fragmentation.

Notes

1. Map by Miriam Hamilton. © 2025 National Center for State Courts.
2. Shwarnali Bhattacharjee (University of North Texas) and Hasan Ahmed (University of Maryland, Baltimore County), The Sundarbans: A Gift to Bengal.
3. Colantonio A., B. Moldofsky, M. Escobar, L. Vernich, M. Chipman, and B. McLellan. "Using geographical information systems mapping to identify areas presenting high risk for traumatic brain injury." Emerg Themes Epidemiol. November 4, 2011, 8:7. doi: 10.1186/1742-7622-8-7. PMID: 22054220; PMCID: PMC3260231.

References

African People & Wildlife: https://www.africanpeoplewildlife.org/
Air quality: https://storymaps.arcgis.com/stories/5071792639ef47729fad54da835d37d3
Alcis: https://www.alcis.org/
Amazon Conservation Team: https://www.amazonteam.org/
Blue Water GIS: https://bluewatergis.com
Boundary markers: https://storymaps.arcgis.com/stories
    /920e34eb52dc41c5acf9b626af2d0c1b
Brain trauma: https://pmc.ncbi.nlm.nih.gov/articles/PMC3260231/
Charles River Watershed Association: https://www.crwa.org/
City of Tucson Equity Priority Index: https://teds.tucsonaz.gov/pages/index
Clean Air Carolina: https://cleanairenc.org/
Climate change preparation: https://storymaps.arcgis.com/stories
    /ade9358709ab4af39c7fc5de610279eb
Confederated Salish and Kootenai Tribes: https://cskt.org/

COVID 19's impact: https://storymaps.arcgis.com/collections
    /74981ffa579e4267bbbf66d488bb38fc?item=39
Dam removal: https://storymaps.arcgis.com/stories/62917edcb76c4e10868cbb7a79638282
Delaware Geological Survey: https://www.dgs.udel.edu/sites/default/files/publications
    /info6.pdf
Dena Kayeh Institute: https://denakayeh.com/
Electric commuting: https://storymaps.arcgis.com/stories
    /8b83356c064b4663bfb4b04dbe8677eb
Equity priorities: https://storymaps.arcgis.com/stories
    /9a0ab238488240d6b4fad20a50b1d795
Esri: https://www.esri.com
Historic maps: https://storymaps.arcgis.com/stories/ffb54e057a654655a73d3d2b6f8a7ae6
Human-elephant conflicts: https://storymaps.arcgis.com/stories
    /6cf68501c25f470fbf27faf2a84f704d
Hurricane preparation: https://storymaps.arcgis.com/stories
    /d3052f807a6343709104f3232259b992
Indigenous environmentalism: https://storymaps.arcgis.com/collections
    /1551802e8e8c4a1d9f3bde7bc9bba1aa
Indigenous knowledge: https://storymaps.arcgis.com/stories
    /d2b6d7f98fbb4ed1a93aff91bac1b6be
Institute on the Environment: https://environment.umn.edu
Latinos in Heritage Conservation: https://www.latinoheritage.us/
Legal deserts: https://experience.arcgis.com/experience
    /832501b9ffe74b21a79a5a3910d7f7e7/page/Home
Mapping hunger: https://storymaps.arcgis.com/stories
    /ca3867611cbc4547a8a14ecaa06d8161
Mapping vaccinations: https://storymaps.arcgis.com/stories
    /cb81725576154ddbbdc5d7120de58a68
Médecins Sans Frontières: https://www.msf.org
Migrating farmworkers: https://storymaps.arcgis.com/stories
    /4a2e09dcc5474a12a34993ec8eb06620
Mine maps: https://storymaps.arcgis.com/stories/4586c60dc91744cbae9967442f990468
National Center for State Courts: https://www.ncsc.org
Nature's diversity: https://storymaps.arcgis.com/stories
    /c633976de1d1492c86c99dcb6473cf13
Nelson City Council: https://www.nelson.govt.nz
New York University Center for Urban Science + Progress: https://engineering.nyu.edu
    /research/centers/cusp
Nomadic tribes: https://storymaps.arcgis.com/stories
    /faf39b39ad0640fd895733207173ab56
Nutrient diversity: https://umn.maps.arcgis.com/apps/Cascade/index.html?appid=
    a48c26df4577490ba8b92d410df2e1fd
Omacha Foundation: https://omacha.org/
Open space: https://storymaps.arcgis.com/stories/76d3f217f5974188a41b81434eb3cca2
Open Space Institute: https://www.openspaceinstitute.org/
Port positioning: https://storymaps.arcgis.com/stories
    /43025cdcf7c5460094d0169a0d99714a
Prague Institute for Planning and Development: https://iprpraha.cz/en/

Racial segregation: https://storymaps.arcgis.com/stories
/bd15a5eb9eae49cda09bfa7368272f89
Research centers: https://www.esri.com/en-us/about/about-esri/international-rd-centers
River dolphins: https://storymaps.arcgis.com/stories
/d995fb44a447430295786a51c3253120
River ecosystems: https://storymaps.arcgis.com/stories
/df468704609b472f846330f84b42334b
S&P Global: https://www.spglobal.com
San Juan Island National Historical Park: https://www.nps.gov/sajh
Seeds for Development: https://www.seedsfordevelopment.org
Stanford University Libraries: https://library.stanford.edu/
Territorial justice: https://storymaps.arcgis.com/stories
/6fa6d4807842404187981cc218bf4394
The Harvard Chan National Institute of Environmental Health Sciences (NIEHS) Center for
Environmental Health: https://hsph.harvard.edu/research/environmental-health-niehs/
The Sundarbans: https://storymaps.arcgis.com/stories
/08fe78971a83403d859f2a18eb413228
UNHCR, The UN Refugee Agency: https://www.unhcr.org
University of Arizona Libraries Data Cooperative: https://lib.arizona.edu/research/data
/geospatial
Urban diversity: https://storymaps.arcgis.com/stories
/3ed83ef359d346618510764c6222ac01
USAID, Bureau of Humanitarian Assistance: https://www.usaid.gov/

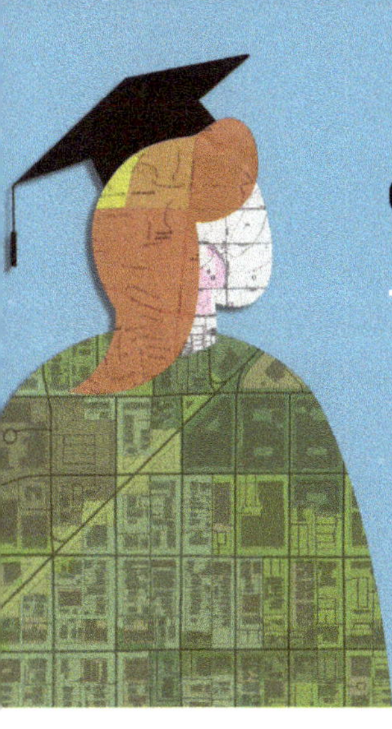

CHAPTER 3

Teaching

Introduction

Geospatial tools typically play two roles in a higher education curriculum. One role involves the teaching of students who intend to become GIS professionals, whether through a GIS certificate program or a bachelor's or master's degree program in geographic information science. The other role involves the use of GIS and related software, such as ArcGIS StoryMaps and ArcGIS Survey123, to help students in a range of fields understand the spatial aspects of their discipline or a particular classroom assignment. This chapter will discuss both roles and show the work of students in both tracks.

The GIS professional curriculum

The typical GIS curriculum for students intent on becoming GIS professionals offers courses in three broad areas:

- **Required** courses that provide both conceptual and theoretical underpinnings and practical and technical skills for a comprehensive understanding of GIS
- **Elective** courses that enable students to focus on technical or conceptual courses matching their interests and career goals
- **Outside** courses that give students opportunities to connect GIS to other areas of interest, with options for a graduate minor in a related field

Key components of a professional GIS curriculum include the following:

- **Core courses:** foundational knowledge in GIS theory, spatial statistics, remote sensing, and research methods, often including an introductory seminar
- **Technology and methods courses:** focus on specific software (for example, ArcGIS), programming (Python for GIS), databases, and technical skills such as geoprocessing, cartography, and data visualization

- **Application and elective courses:** specialization in areas such as environmental science, public health, urban planning, business analytics, or digital archaeology, using GIS for real-world problems
- **Professional practice courses:** including projects, theses, or internships, emphasizing ethical considerations, data management, and communication

The competencies gained in a typical GIS curriculum include an understanding of geospatial concepts and methods, an ability to analyze spatial data, a skill in cartographic and visual communication, a capacity to manage spatial data, a knowledge of relevant program languages, a familiarity with statistical analysis, and a capability to manage projects. The courses typically offered in a GIS curriculum include the following:

- **Introduction to geographic information systems:** an overview of core GIS concepts, principles, data models (vector and raster), data manipulation, and basic spatial analysis techniques, as a prerequisite for more advanced courses
- **Spatial analysis/geographic information analysis:** the methods and tools for analyzing spatial data to solve geographic problems, including topics such as point pattern analysis, overlay analysis, interpolation, and surface analysis
- **Remote sensing and image processing:** the principles of acquiring, processing, and analyzing remotely sensed data (satellite imagery, aerial photography, lidar, and so on) for various applications
- **Cartography and visualization:** the theory and practice of map design and effective visual communication of geographic information, using traditional and modern web-based tools
- **GIS programming and software development:** the use of programming languages, most commonly Python, to automate workflows, customize software functionality, and manage geospatial tasks
- **Spatial database management/geodatabases:** the design, creation, management, and retrieval of geospatial data using database systems such as SQL and PostGIS
- **Geospatial statistics/applied spatial statistics:** the statistical foundation for analyzing spatial data and understanding spatial patterns and processes
- **Geospatial systems analysis and design/project management:** the systematic approach to designing and implementing complex geospatial information systems and managing projects effectively
- **Capstone project or master's project:** a culminating experience in which students apply their accumulated knowledge and skills to a real-world problem, often in collaboration with an external partner

The curriculum for an MGIS degree requires two years of full-time study, although some programs have shortened the time to one or one and a half years by going year-round, using online course delivery, having cohort models, or allowing higher course loads. A full-time student would normally take three to four courses each semester. Shorter GIS certificate programs also exist that allow students to reduce the cost of education and move into employment more quickly, although the scope and depth of student knowledge will necessarily be more limited. A certificate program, for example, might take nine months, be entirely online, and have project-based courses aimed at preparing students to become GIS technicians, who, with some experience, can then become GIS specialists and eventually a certified GIS analyst.

That certification process involves a few steps:

- Students, recent graduates, and mid-career users just entering the field can get a GIS Certification Institute (GISCI) GISP-E or pre-GISP credential.
- Esri offers technical certifications at the foundation, associate, and professional levels for users who can validate their expertise in ArcGIS.
- The main certification for GIS professionals is the GISP, which requires a portfolio that documents one's GIS education, experience, and contributions to the field. It also requires passing the proctored, twice-yearly GISCI Geospatial Core Technical Exam, which tests knowledge of core GIS concepts and understanding of the professional code of conduct.

The GIS-assisted curriculum

For students not intending to become GIS professionals but who use geospatial tools in their academic work, story maps offer one of the best options. The following curriculum draws from the work of the University of Minnesota Story Maps Curriculum Team, whose work is useful for students and instructors in teaching and learning about GIS. Although many students have been exposed to digital maps, instructors can't assume that every student has experience with GIS, so story maps provide a template for laying out text and images while enabling students to explore the spatial aspects of a topic. The ability of students to publish their story maps allows them to share their work with the public, making their research more accessible.

When creating an assignment, instructors might ask the following questions:

- How does spatial thinking enhance the study of the course's topic? Make it clear to students that a spatial lens can enhance their understanding of the course material. The maps they produce should tell a story or advance an argument.
- What do you want students to map? Choose topics that lend themselves to spatial analysis, with locations that students can easily find and map and where space and place play an important role.
- What sources should students refer to for their maps? Encourage students to work with texts relevant to the course, whether it's a book chapter, newspaper article, or any source that discusses locations.

- What kinds of relevant data are available to students? Spatial data includes geolocated points and layers, which show concentrations of activity and relationships among places. Students should look for existing data—such as political boundaries, demographics, or topography. As the instructor, you should search for potential data before students start the assignment from sources such as ArcGIS Living Atlas of the World or census data.
- What else should students be aware of? When mapping historical sites, students should be aware that historical and modern places may not align and such locations may need to be approximate. Story maps don't have a simple way of including citations, so consider creating a citations template for your students ahead of time to make sure that they properly recognize sources.
- What copyright restrictions exist? When bringing in external media, especially images, make sure that students do not violate copyright and encourage them to use open-source websites for images when possible.
- Should this project be public facing? Students may want their projects publicly accessible, which can motivate them to make their maps more presentable. Do not distribute their work without their knowledge and consent.
- What might confuse or be challenging for students? Although most students find it easy to use a story map app such as ArcGIS StoryMaps, ArcGIS Online takes more time to learn. ArcGIS StoryMaps and ArcGIS Online are connected—but separate—apps. ArcGIS Online is the mapping software as a service within which students can plot points, perform analyses, and build a web map. ArcGIS StoryMaps is an online digital storytelling app into which you can load web maps and other types of multimedia to display and share a narrative presentation (a "story").

ArcGIS Online and ArcGIS StoryMaps compared

	ArcGIS Online	**ArcGIS StoryMaps**
What is it?	The online mapping software as a service used to build web maps. Students upload data into ArcGIS Online and design and build maps.	The online web app in which students can complement their maps with text, pop-ups, points, and multimedia in an interactive and compellingly designed narrative.
What content can it include?	Students can add any existing data, upload their own CSV files, or include map notes for additional features.	It can feature web maps that students and others create but also text and multimedia—video, photos, audio, hyperlinks, and so on.
What sort of customization is available?	Students can adjust the symbology of points and features and choose an appropriate basemap.	On the side panel of each story, students can edit just as they would in a standard word processor. On the main stage, students can toggle map features on and off and upload numerous types of media.

Rubric

A rubric for grading depends on the goals of the instructor. This table provides a suggested rubric, adjustable to fit different course needs.

Sample rubric

Requirements	3: Excellent	2: Adequate	1: Insufficient
The final story map includes multiple locations.			
Each location has an appropriate image and narrative description.			
Each location is relevant to the chosen topic.			
The importance of space and geography to the topic is clear.			
The map and text are well edited and flow together well.			
The story map follows proper citation format and gives photo credit.			
The story map is shared with the instructor and class.			

Assignments for a one- to two-week project

Week 1: Introduction and in-class workshop

Introduction to story maps. Take half a class period for a quick demonstration of story maps, introducing students to what they can do, showing how to set up their sign-in (organizational) accounts, and walking them through the basic steps of building a map. Hand out the assignment prompt.

Set up class group. Once students have created their accounts, create a class group for them to share their projects and for you to view and grade their projects. Invite every student in the class to this group.

In-class workshop. Over an entire class period, let students work on their maps and ask questions. Ask your students to create a map tour before class so they become familiar with the application and find all their locations and relevant images. Review some basic steps, such as linking in images, searching for locations, or dropping pins, with the goal of students mapping all their points by the end of the class period.

Week 2: Submission of final map

Finished story map. For their final project, students should complete a story map with at least eight mapped points with descriptions and images for each.

Assignments for a four- to six-week project

For projects that last several weeks, we recommend breaking them into four major parts: a brief introduction to the program, a storyboard outline, an in-class workshop, and a finished story map project. Students should have multiple steps for each component. The following list is an example of a four-week assignment.

Week 1: Introduction to story maps

Reserve half a class period for a quick presentation and demonstration of story maps. Use this time to introduce students to what story maps can do, familiarize them with the ArcGIS StoryMaps website, set up their sign-in accounts, and walk them through the basic steps of building a map in the application. Select some relevant story maps before class; you can find examples in the ArcGIS StoryMaps Gallery at doc.arcgis.com/en/arcgis-storymaps/gallery. Hand out the assignment prompt the first week.

Week 2: Storyboard outline

A storyboard describes the proposed subject, locations, sources, and text of each section of the story map, providing students with structure for their project as well as a setting of points to start mapping. To give students enough time to respond to feedback, grade this work and return it to them within a week.

With the storyboard, students should write a paragraph on the intended topic, thesis, and source approach of their mapping project. They should include a short bibliography to show that they have thought through the primary and secondary sources for their work.

Week 3: In-class workshop

Dedicate an entire class period for students to work on their story maps and ask questions. Require students to begin creating a story map before class so that they become familiar with the user interface and functionality and encourage them to select a template, start a map, and write relevant text. Students should have all their points mapped by the end of this class period, and they should begin to add text and other media.

Week 4: Finished story map

Students should submit, as their final project, a complete story map, with at least six points and eight sections or panels. Further requirements are listed in the following list for a full-term project. Have students share their work by email or as a link to whatever learning management system you use.

Assignments for a full-term, 15-week project

Digitally based projects require a lot of scaffolding, so have components spread throughout the semester to ensure that projects run smoothly. The following list is a timeline for a 15-week semester, with detailed descriptions of each step and suggestions for how steps can be implemented.

Weeks 2–3: "Introduction to story maps" presentation

Reserve one of your class periods to give a presentation and demonstration of story maps and use this time to introduce students to spatial thinking and the ArcGIS StoryMaps website. Set up student sign-in accounts and walk through the basic steps of building a map in the program. Select some relevant story maps as examples before class from the ArcGIS StoryMaps Gallery.

Weeks 5–6: Topic proposal and bibliography

Have your students write and submit a brief paragraph describing their chosen topic and a central question they will address. Students should describe how they will incorporate spatial analysis into their assignment, proposing some of the locations they may focus on and the sources they intend to use to determine points. They should also include a short bibliography to show that their topic has viable and reliable source options.

Weeks 7–8: In-class workshop and storyboard

Toward the middle of the semester, dedicate one of your class periods to walking students through the data creation and mapping process. Students should come to class with two items: a CSV spreadsheet file and a storyboard (a tabled outline that describes the proposed subject, locations, and text of each section of the story map). This will provide students with some structure for their project as well as a set of points to start mapping in class; these deliverables should be graded and returned within a week.

Students should create a CSV file spreadsheet of the points they plan to map. This should be created in Microsoft Excel or Google Sheets but must be saved as a CSV file. This need not be complete but should include the name of the place, its GPS coordinates, and a description of the place. In class, students will then upload their spreadsheet to ArcGIS Online and start adjusting their points to create a map.

Weeks 12–13: Full draft of map

Before the final assignment is due, students should turn in the first draft of their story map. The essential components for this draft include having all points mapped and complete text in the side panel. Stylistic editing is less important at this step, but all the content should be present. Promptly grade these drafts and return them to students to make final revisions, stylistic edits, and work out any kinks with the application. Grade the draft on its content, research, argument, and the effectiveness of its maps with symbology, scale, and relevance of points.

Week 14: Workday (*optional*)

You may want to dedicate one of your last few class periods to a workday for students to finalize their projects by having students ask technical questions and share feedback.

Week 15 (finals week): Final story map

This final draft should be shared by email or as a link in whatever learning management system you use. You may want students to present their maps to the rest of the class or pick examples to look at together. These presentations can be short, and students should not attempt to cover all content in their maps.

Grading

Grading story maps depends on your goals; you might want to place more emphasis on the visual components of the map instead of its textual content, or vice versa. You also might adjust the grades based on what matters most for the class, although students should receive a grade for two components: the storyboard and the story map.

For a one- to two-week project, the story map might be worth between 10 and 15 percent of the final grade. For a four- to six-week project, it might be more like 20 to 25 percent of the final grade. Of that grade, the storyboard might be 5 percent, based on the viability of the proposed story map, the thoroughness of the storyboard outline, and the completeness of the bibliography. The finished story map might be 15 to 20 percent. For a full-term project, the story map might be 40 percent or more of the final grade. Within that, the topic proposal and bibliography might be 5 percent, based on completion and effort; the in-class workshop might be 5 percent, based on attendance at the workshop, spreadsheet completeness, and the quality of the storyboard; the first map draft might be 5 percent, and the final story map, 25 percent, based on the preceding list.

Group assignments

Most of the preceding information applies to group projects, as well, with a few exceptions. Before handing out the assignment, you will need to create groups in ArcGIS Online where students can add content. Each student group should have its own ArcGIS Online group, with the instructor as the group owner. Groups should be between three and four people, depending on the class size, and you should place students in groups before handing out the assignment.

The rest of this chapter will focus on a range of examples of student work, some of it from capstone projects of GIS students and some from students using GIS in coursework in other disciplines. What these examples show is the extent to which GIS has become pervasive not only in the work of faculty and staff, as other chapters in this book illustrate, but also in the work of undergraduate and graduate students in a remarkable range of disciplines.

Arts examples

Mapping Dante's *Inferno*

DISCIPLINE Literature

ORGANIZATION Student

INDUSTRY Education

LOCATION Global

World literature is full of fictional geography, such as Dante's *Inferno*, with its descriptions of the circles of hell. This story map, made by a Czech cartography student for his master's thesis, traces the otherworldly geography of this 14th-century epic poem onto a map of medieval Europe, based on both the author's intensive descriptions and the calculations of Antonio Manetti and Galileo Galilei. The latter mistook the size of the earth and the distances between landscape elements, but Galileo and Manetti's map of *Inferno* is critiqued and corrected in this story map, based on our current geographic knowledge. The author of this story map also created a 3D model of to show what Dante—and Manetti and Galileo—had in mind when he envisioned the circles of hell. This project shows the power of what happens when we map works of the imagination, turning verbal descriptions into visual documents of great appeal.

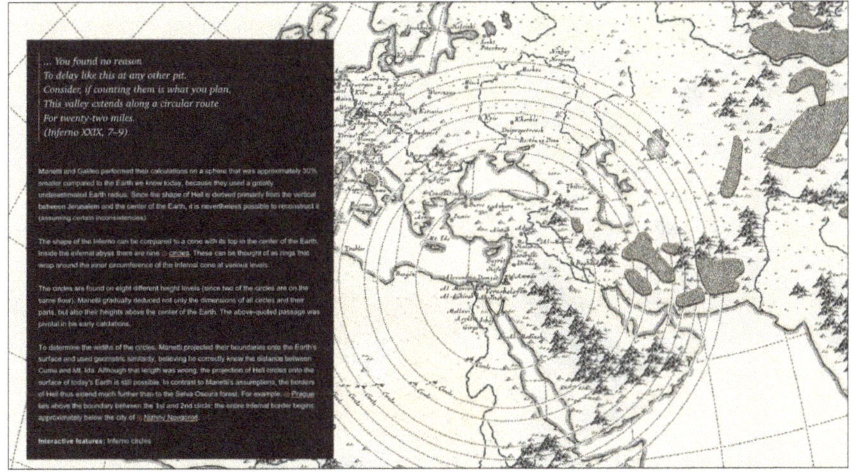

This map visualizes the circles of hell in Dante's *Inferno*, allowing readers to imagine a literary underworld. *Copyright © 2022 Josef Münzberger.*

Mapping James Joyce's *Ulysses*

DISCIPLINE Literature   **INDUSTRY** Education

ORGANIZATION Student   **LOCATION** Dublin, Ireland

Maps have informed literature from its very origins in the West, evident in the vivid geographic descriptions in the ancient Greek epic poems *Iliad* and *Odyssey*. The emerging field of literary geography now considers everything from the imaginary geographies in works of science fiction to the actual geographies in works of historical fiction. James Joyce's modernist parallel of Homeric poetry, *Ulysses*, draws ideas and themes from the classical *Odyssey* and features numerous references to Dublin, Ireland, where Joyce was born. This story map visualizes city locations in the novel; the page numbers in the book are linked to map locations where events in the novel take place over the course of a day: June 16, 1904 (known as Bloomsday after the protagonist, Leopold Bloom). Such maps allow readers to see the sites that influenced their favorite novelists. The story map illustrates how space and time help shape human imagination, turning fact into fiction and back into fact.

American jazz clubs

DISCIPLINE History   **INDUSTRY** Education

ORGANIZATION Academic   **LOCATION** St Louis, Missouri, USA

History reminds us of what *was*, allowing us to imagine what *could be*. This story map by a student in heritage studies and public history traces the history of jazz clubs in three American cities—St. Louis, Minneapolis, and Detroit—and tells the story of the music played in those cities. It also reveals the appalling record of urban renewal, which led to the demolition of the major jazz clubs in those cities. As the story map puts it, "The buildings that once struggled to contain the music as it seeped through the cracks in the mortar have been replaced by highways, university campuses, and the sterile facades of downtown development. What remains are memories, mentions in newspaper archives, and the rare recording." The narrative reminds us that "The venues may be lost, but music is not."

Mapping temporary memorials

DISCIPLINE Arts

ORGANIZATION Academic

INDUSTRY Education

LOCATION Minneapolis, Minnesota, USA

We typically think of memorials as permanent things, but the temporary ones that spring up where tragedy has occurred in a community can have an even more powerful effect on people's memories and immediate responses. The story map that a graduate student created as part of a capstone project locates the "spontaneous shrines" that mark where young people died in Minneapolis, mainly from gun violence or auto accidents. She not only maps the shrines but also includes the names of the victims and photographs of the memorials, whose impromptu nature means that they will not last long. The story map also includes links to the data on accidents and crimes and information about the organizations devoted to humanizing the statistics. Efforts like this can make these temporary memorials a permanent part of a community.

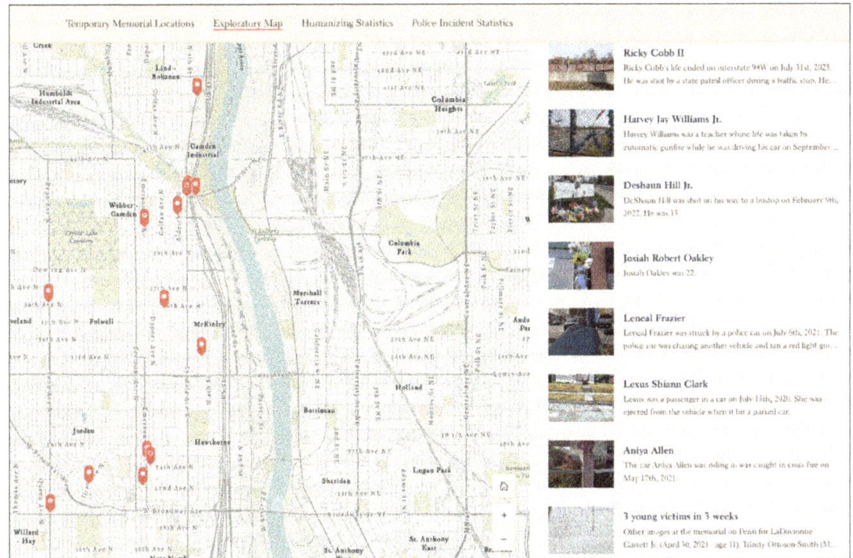

This map shows the sites of temporary memorials to young people killed in automobile accidents or gun violence in Minneapolis. *Courtesy of Margarette Nevalainen.*

Painting with GIS

DISCIPLINE Arts　　**INDUSTRY** Education

ORGANIZATION Student　　**LOCATION** Global

Artists and mapmakers may seem far removed from one another, but the relationship of maps to depictions of the environment goes back to some of the earliest cave paintings. Artists also like to explore new mediums for their expressive possibilities, as Wes Jones has done here, using GIS software to create a landscape painting that has the vertical orientation of a Japanese screen print and the minimalist, monochromatic quality of a Chinese ink painting. Jones created the scene by combining the symbology and other properties of ArcGIS Pro with scanned images of hand-painted ink washes, producing a work that blurs the boundaries between geography and landscape painting. There's an art to making maps—and there's a maplike way of creating art.

Wes Jones has used ArcGIS Pro as a creative tool to reimagine landscape art reminiscent of Japanese *sumi-e* and Chinese *shui mo hua* ink paintings. *Courtesy of Wes Jones.*

Geolocating Atlanta's rap

DISCIPLINE Music

INDUSTRY Education

ORGANIZATION Student

LOCATION Atlanta, Georgia, USA

Music is both acoustic and spatial, and some genres of music are also highly geographic. In hip-hop, for example, musicians make frequent references to particular places. Born out of the struggles of people living in poverty, hip-hop and rap often speak to the places that musicians know best, as shown in this map of locations from Atlanta-area song lyrics. The map identifies places in and around the city mentioned by rappers and hip-hop artists (listed on the left), and each point is linked to the relevant lyric. This is an example of counter-mapping—documenting the geography of marginalized groups, alternative narratives, and overlooked histories rarely captured in maps made by the majority demographic. Counter-mapping seems especially relevant to hip-hop, which runs "counter" to earlier, more traditional musical genres.

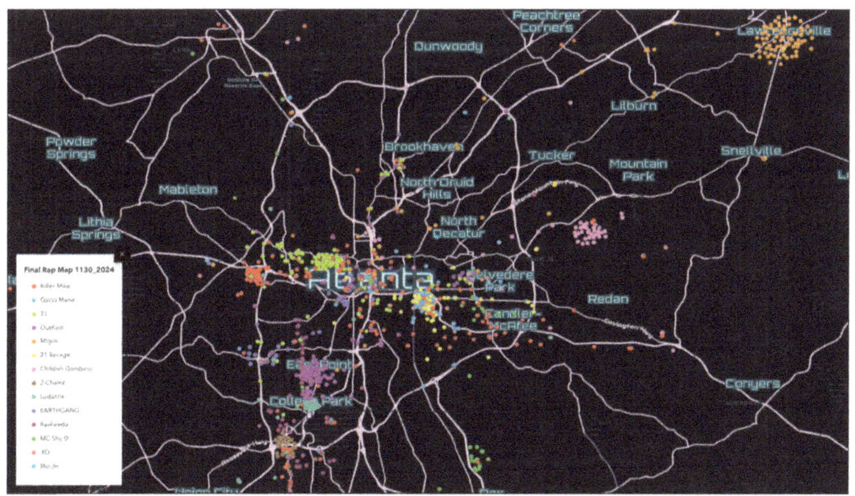

This map of Atlanta shows locations that local hip-hop musicians refer to in their lyrics.

Courtesy of the Atlanta Rap Map, Georgia State University.

History examples

Preserving Native American sites

DISCIPLINE History

ORGANIZATION Student

INDUSTRY Education

LOCATION Las Vegas, Nevada, USA

This story map by a graduate student at the University of Nevada, Las Vegas, looks at Brownstone Canyon, where there are 42 recorded sites of roasting pits which Native American communities used as "kitchens" to roast local agave plants. The project highlights the threat that a new housing development adjacent to Las Vegas presents to these archaeological sites. It also lays out a future research agenda that includes radiocarbon dating of the roasting pits and searching the sites for artifacts and any organic remains from the roasting process to determine whether the process resulted in the fermentation of agave. The winner of a student award in an Esri story maps competition, this entry shows how maps tell stories and raise awareness about what is happening in a city's backyard.

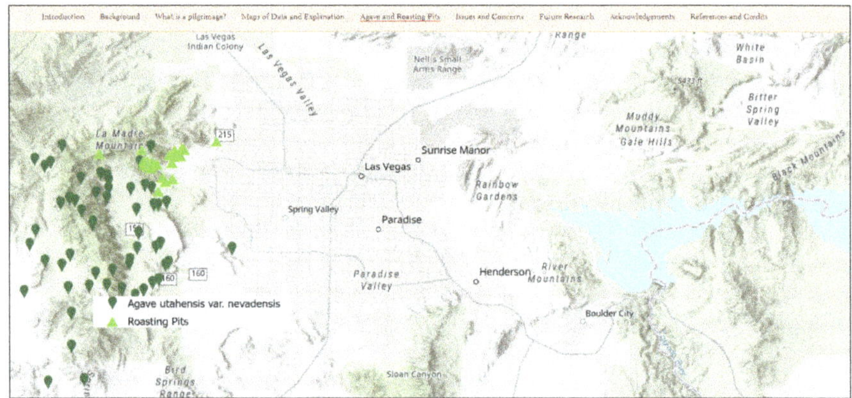

The suburban sprawl of Las Vegas now threatens Native American archaeological sites in the adjacent Le Madre Mountains. *Courtesy of Ruzena Zatko.*

Enslaved peoples' migration

DISCIPLINE History   **INDUSTRY** Education

ORGANIZATION University of Richmond   **LOCATION** USA

Maps can reveal geographies that we have either forgotten or may have historically wanted to hide. This story map of the movement of enslaved people in the US in the early 1800s shows how plantation owners moved slaves from the Eastern Seaboard states to the Deep South to pick cotton, without pay. At the height of the cotton boom in the 1830s, almost 300,000 slaves were forcibly moved from the areas in blue to those in red. The story map includes a decade-by-decade map of this forced migration as well as excerpts from the stories that enslaved people told about their experiences. This is history that every American should know and never forget.

Mortality in medieval Europe

DISCIPLINE History   **INDUSTRY** Education

ORGANIZATION Student   **LOCATION** Italy

Rural communities are often overlooked in standard histories, which is especially true in medieval Europe, where literacy was less common in rural areas than in cities. This story map tracks the spread of the plague epidemic in 14th-century Europe, starting at major Mediterranean ports and moving inland along trading routes, an epidemic that resulted in deserted villages, abandoned fields, mass graves, and untended animals. Port cities such as Venice suffered the most; there, 70 percent of the population died, as did 20 of the 24 leading physicians. And as people left cities to seek refuge in rural areas, they often brought the plague with them or contracted it from abandoned goods they found there. The power of this story map is not just its compelling narrative but the maps and images that make it real to readers.

Nightlife in old Shanghai

DISCIPLINE History

INDUSTRY Education

ORGANIZATION University of Minnesota

LOCATION Shanghai, Shanghai Municipality, China

Maps can show relationships that historical records may not reveal. That became apparent with a six-week student exercise in which student groups made story maps about various aspects of 20th-century Shanghai, mapping its nightlife, public spaces, and the locations of crime, among other data. As part of a research methods course, the students learned how to geocode to turn text into mappable data points, drawing from primary sources. They also mapped information without knowing, at first, what connections among the data they would find and then created their story maps based on the relationships that emerged from their work. One story map, for example, addressed the spatial proximity that existed between Shanghai's police stations, dance halls, and prostitution activity, based on police records from 1920 to 1947. Maps can show that activities occurring close to one another may be more intricately connected than we think.

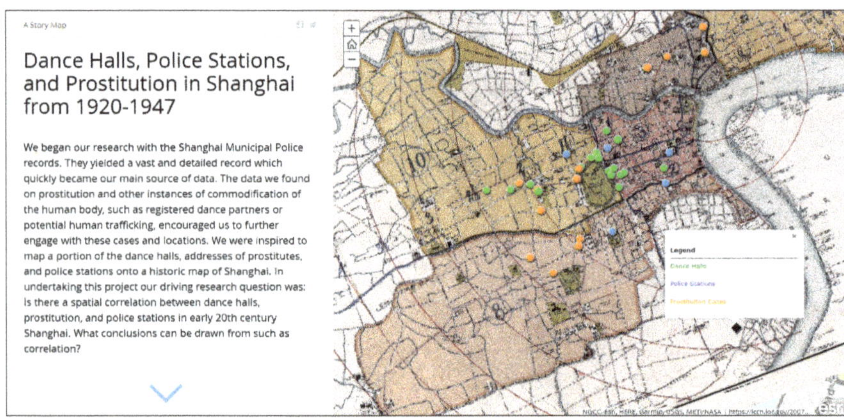

This map of police stations, dance halls, and prostitution activity in mid-20th century Shanghai shows how spatially connected these activities were. *Courtesy of Dr. Christopher Saladin.*

Spanish silver trade

DISCIPLINE History

INDUSTRY Education

ORGANIZATION University of Minnesota

LOCATION Bolivia

Short-term group projects can achieve a lot in a little time, as happened with this group project in a history class where students made a story map based on an in-class group exercise on the early silver trade in Latin America. The students each focused on different aspects of the topic, drawing from the course's text and additional research, and identified the relevant locations on their ArcGIS Online map. They then described their findings of the impact of the silver trade on the Indigenous population and landscape in a story map. The students noted how the mapping exercise changed their views of each topic, spatializing information in their readings. One student observed that visualizing the movement of silver even raised questions about the route of the silver supply described in their textbook, which didn't account for the rugged terrain in their area. Words and maps both have a role to play in our understanding, but maybe when it comes to understanding trade, it's a trade-off.

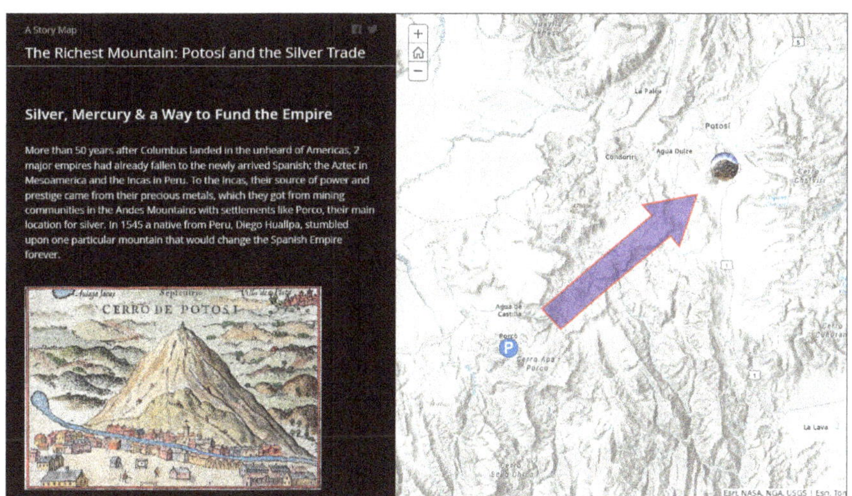

Maps can enhance our understanding of historical events, especially when they involve the movement of people and goods, as in the silver trade. *Courtesy of Dr. Christopher Saladin.*

Politics and government examples

Ownership of Indigenous land

DISCIPLINE Journalism

ORGANIZATION Student

INDUSTRY Local Government

LOCATION Minneapolis, Minnesota, USA

Journalists tell stories about people, places, and events relevant to the present, and this story map by a journalism student illustrates the tremendous change taking place at the highest falls of the Mississippi River, which the Dakota people called *Ȟaȟa Wakpá*. The US government has disposed of the Upper St. Anthony Falls lock and the adjacent land, and a Dakota-led nonprofit, Owámniyomni Okhódayapi, has received 5.2 acres from the government, with another three acres to come, to restore the ecology of the riverfront land and honor Dakota history and culture. This reclamation includes remembering Wíta Wanáǧi, or Spirit Island, which once stood in the river and where Dakota people would go to give birth in the mist coming off the falls. The lock is now closed to boat traffic, and a study is under way to explore the feasibility of its removal. This is a story with a long history and a promising future.

The Dakota people will once again have ownership of some of their sacred land adjacent to the highest falls of the Mississippi River. *Courtesy of Jasmine Shackleford.*

The culture-nature divide

DISCIPLINE Urban Planning

ORGANIZATION Student

INDUSTRY Local Government

LOCATION Minneapolis, Minnesota, USA

Geographic proximity to a park doesn't always mean that the people in an adjacent neighborhood feel welcome there. The location of Theodore Wirth Park next to north Minneapolis highlights the challenges of making a community–park connection. This story map, by a graduate student, collaborating with the Center for Urban and Regional Affairs and NatureWerks, explores the history of the diverse population in north Minneapolis and the reasons that some people living there haven't felt welcome in the nearby 759-acre park. The city has one of the largest gaps in homeownership rates and educational achievement between Black and White residents in the nation. An affluent population on one side of the park and a less affluent population on the other made the latter feel as if they didn't belong in the park. The story map captures the stories of people living in north Minneapolis and explores how they could feel more at home in the park, as well as how frequent users of the park could feel more welcome in the adjacent neighborhood commercial corridors. As this story map shows, streets go two ways.

Femicide in Türkiye

DISCIPLINE Sociology

ORGANIZATION Student

INDUSTRY Local Government

LOCATION Türkiye

Femicide remains a real problem in places such as Türkiye, where four in 10 women experience physical or sexual violence. This story map, by three university students in the United States, tracks the gains that women made in the early 20th century in Türkiye and discusses the Istanbul Convention on combatting violence against women. They also map the locations in Türkiye where a substantial number of women have been killed at the hands of their intimate partners, with profiles of some of these women to give a human face to the descriptions of how they died. Although Türkiye has withdrawn from the Istanbul Convention, this story map calls on the government to address an issue on which it once led the international community.

Shared feelings about places

DISCIPLINE Urban Planning

ORGANIZATION Student

INDUSTRY Local Government

LOCATION Twin Cities, Minnesota, USA

Digital information can help us understand social phenomena in new ways, and this story map by two graduate students makes that case. They looked at Airbnb listing descriptions in several cities in North America and Europe, researching the thoughts and feelings about those cities that customers expressed in their reviews. The researchers then grouped similar wording in these reviews to identify shared urban characteristics of each city. By scanning 687,000 listings in 176 cities, the two students found that some cities, such as New York City and Boston, have many shared characteristics that they don't share with nearby cities, such as Washington, DC, and Baltimore. They found the same in Europe, with Rome and Milan sharing traits that they don't share with Naples. What we say online about our localities may say more than we think about ourselves and the places we live.

Airbnb online reviews offer insights into how people feel about the cities in which they have stayed, providing information about our emotional reactions to places. *Copyright © 2024 Jina Kim and Yao-Yi Chiang.*

African American golfing

DISCIPLINE Athletics

ORGANIZATION Student

INDUSTRY Local Government

LOCATION Minneapolis, Minnesota, USA

Few golf courses are included in the National Register of Historic Places, but one of them, the Hiawatha Golf Course in Minneapolis, has played a major role in the African American community since the 1930s. Located on wetland next to Lake Hiawatha, the golf course has hosted golf tournaments and provided a welcoming place for the community to socialize and compete in the sport. This story map recounts that sports history and takes on the conflict that has arisen from a new master plan, approved by the Minneapolis Park and Recreation Board, that would reduce the 18-hole course to nine holes and restore some of the original wetland to reduce flooding in the adjacent residential neighborhood. The African American community has raised concerns about the plan, which highlights the competing needs that can arise between environmental and historical interests.

Public health examples

Pediatric malaria in Nigeria

DISCIPLINE Public Health

ORGANIZATION Student

INDUSTRY Health and Human Services

LOCATION Nigeria

Malaria, caused by the bite of infected female mosquitoes, is epidemic in Nigeria, which in 2020 accounted for 27 percent of global malaria cases and 32 percent of malaria deaths. Children are especially vulnerable; about 300,000 children die of the disease in Nigeria each year. This story map by a student at the University of Ibadan maps the malaria hot spots in the country and spots where malaria is less prevalent—areas that also reflect where childhood poverty rates are the highest and the lowest. The story map makes a case for addressing poverty and the inadequate health care, infrastructure, education, and housing that this data reveals, all of which expose children to the illness and death stemming from malaria.

Mapping medicine's consolidation

DISCIPLINE Public Health   **INDUSTRY** Health and Human Services

ORGANIZATION Student   **LOCATION** Minneapolis, Minnesota, USA

Like many professional fields, medicine has experienced consolidation as smaller hospitals have merged into larger systems and smaller practices have merged into larger groups. This story map by a graduate student shows how that consolidation altered the geography of medicine in Minneapolis. In 1960, seven medical practices were located within walking distance of residents living in census tracts with the highest percentage of Black residents (shown on the left), whereas in 1990, the number had decreased to two. Medical practice not only became concentrated closer to the center of the city and less physically accessible to people in the surrounding neighborhoods, but the number of physicians in large practices also continued to grow; between 50 and 75 percent of Twin Cities doctors work in large health systems—which may not be healthy for everyone.

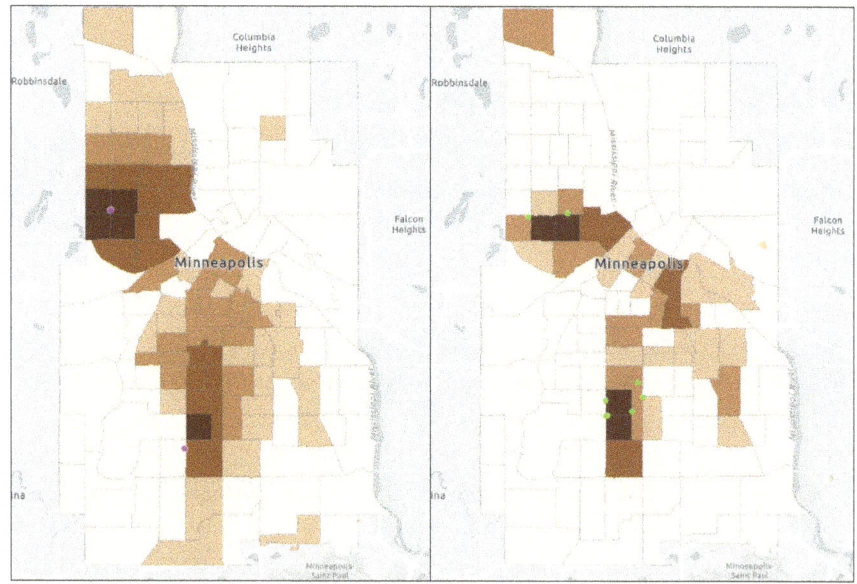

Medical practices have consolidated over the last century, becoming more geographically concentrated. *Courtesy of IPUMS, NHGIS, and Hennepin County Library.*

Mapping mental health

DISCIPLINE Public Health   **INDUSTRY** Health and Human Services

ORGANIZATION Esri   **LOCATION** USA

Psychological data has revealed the decrease in the United States of mental health in many parts of the country. This map[1] shows how dramatically the number of reported poor mental health days rose between 2015 and 2019, before the COVID-19 pandemic, which had an even greater impact on people's mental well-being. This story map on America's mental health decline locates the contributing factors to this situation, including the increase in obesity, unemployment, income inequality, and single-parent households in various parts of the country. Mapping the per capita access to mental health professionals also aligns with the emotional distress of people in particular places, with 47 percent of the population living in areas with a shortage of mental health professionals. Such data reminds us that the first step in addressing a public health problem is to acknowledge it and work to fix it.

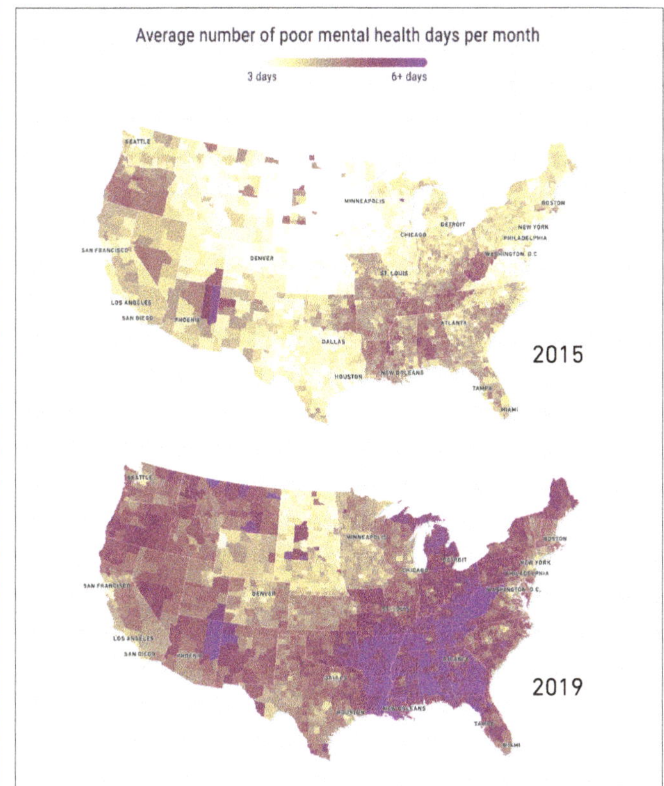

Mental health in the United States has gotten worse since 2015, even as our ability to map the decline has gotten better. *Courtesy of Este Geraghty.*

Bike access in New York City

DISCIPLINE Public Health

ORGANIZATION Student

INDUSTRY Transportation

LOCATION New York, New York, USA

Bike-sharing systems depend on the availability of bikes, especially close to public transit stations to handle first- and last-mile travel. This story map of New York City's CITI Bike program shows the impressive extent of the system, but it also shows where additional stations in key locations would increase bike access and use. The authors adopted a public transport access level (PTAL) metric developed for London's transit system and applied it to New York, showing how it offers a way to assess access to public transportation by diverse populations based on their income, ethnicity, and access to other modes of transportation. This work demonstrates how student work can provide useful, practical, and relevant solutions to the public and public officials.

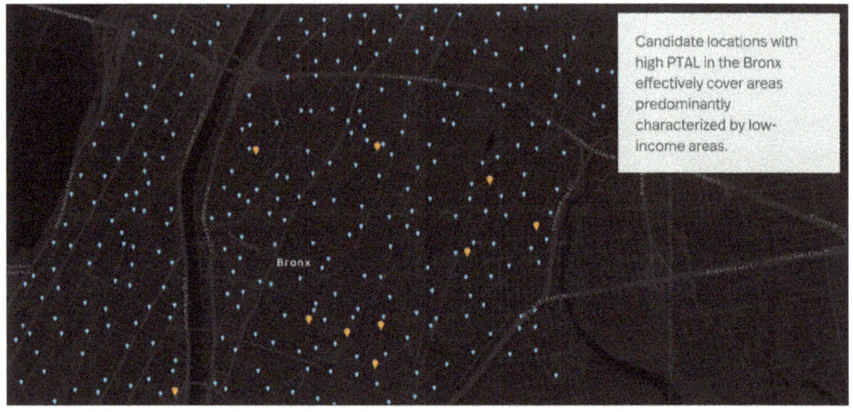

New York City has a sizable bike-sharing system, and this map shows where additional stations would increase access, especially in underserved outer boroughs. *Courtesy of Min Namgung, Yao-Yi Chiang.*

Access to health care

DISCIPLINE Public Health   **INDUSTRY** Health and Human Services

ORGANIZATION Student   **LOCATION** Boston, Massachusetts, USA

The US health-care industry suffers from unequal access to care, depending on several social determinants such as household income, location, and access to transportation. This story map by a Tufts University student in an introductory GIS course looked at various criteria as the basis for selecting the most effective sites for a new community health center in Boston: (a) a site's distance from a transit station and an existing health center, and (b) a location where the local population is at least 12 percent elderly, 20 percent low income, 50 percent ethnic minority, and a medical illness rate of at least 40 percent. Two census tracts fit all these criteria, which led to the identification of two locations that could accommodate a new community health center, best situated to meet the needs of those most in need.

Science and technology examples

Disappearing farmland in the West Bank

DISCIPLINE Agriculture   **INDUSTRY** Education

ORGANIZATION University of Minnesota   **LOCATION** West Bank

The world has followed hostilities in Gaza and the West Bank and the impact of this conflict on cities and settlements, but there has been little coverage of the effect it has had on agriculture and the ability of people to grow food. This study by an undergraduate student looks at changes in land use in the West Bank between 2003 and 2023, showing how, in the blue areas on the map, a once agriculturally rich area in the western half of the territory has stopped food production, which presents an enormous problem for the roughly three million people living in the West Bank. One of the most powerful aspects of GIS is change-detection analysis, highlighting the difference between what was and what is, whether that difference is good or, as in this case, not good.

Tracking bird strikes

DISCIPLINE Wildlife

ORGANIZATION Student

INDUSTRY Conservation

LOCATION Minneapolis, Minnesota, USA

The collision of birds with buildings has contributed to a 25 percent decline in bird populations since the 1970s. This problem is especially acute in locations such as the University of Minnesota's Twin Cities campus, which straddles the Mississippi River, one of the major bird flyways in North America. This story map by a graduate student contains a QR code that provides an online reporting tool for anyone who witnesses a bird–building collision on campus, and it maps the 1,645 collisions that people reported between March 2021 and January 2025. The map also highlights the most problematic buildings, including the university's stadium, neighboring glass-enclosed laboratory buildings, the glass-enclosed pedestrian bridge over the river, and the west bank campus with several glass buildings. The story map documents the bird species most affected, with warblers and sparrows constituting by far the most fatalities, and recommendations for the installation of bird-safe window treatments. It's the only way, as a campus campaign says, to "stop the thud!"

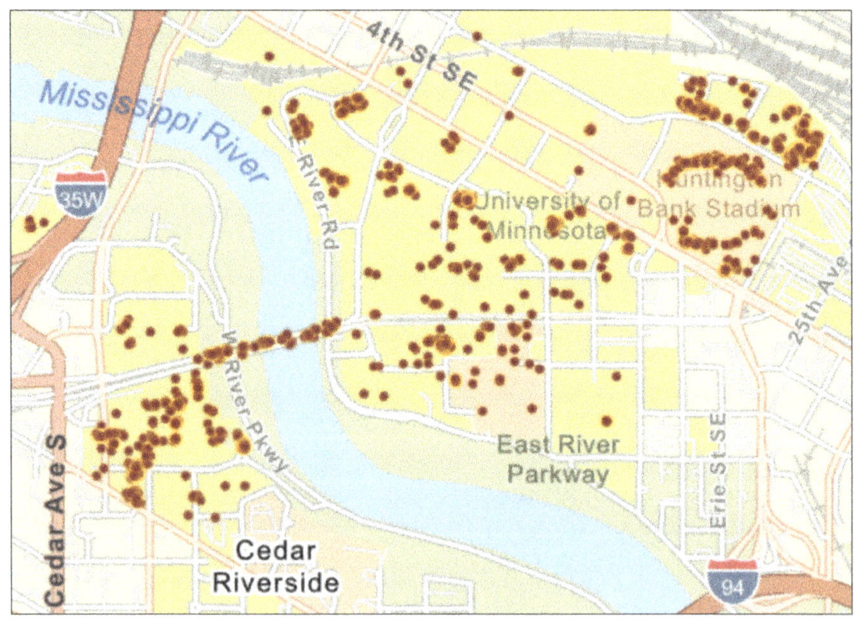

The University of Minnesota straddles the Mississippi River, which is a major flyway in North America, making it a place where birds can collide with campus buildings. *Courtesy of Andrew Hallberg.*

Water quality in farm fields

DISCIPLINE Soil Science

INDUSTRY Water

ORGANIZATION Student

LOCATION Fairmont, Minnesota, USA

Agricultural runoff can pollute the waterways that, in turn, pollute the water sources and recreational lakes that the economies of many rural communities increasingly depend on. This story map by two student interns at the engineering firm ISG recounts how they addressed this problem for Fairmont, Minnesota, where farm runoff was polluting the chain of lakes the city uses for drinking water. ISG designed a "nutrient treatment train" that includes an eight-acre nutrient treatment wetland, with a sediment basin and treatment pond planted with native plants to promote nutrient uptake and provide habitat. The design also features an 8,000-foot-long, two-stage ditch that features inner and outer channels, with vegetated floodplain benches that enhance storage capacity and nutrient uptake while reducing erosion. Together, these elements mitigate nutrient runoff and reduce downstream flooding and erosion, an example of how to engineer a healthier future.

The two-stage ditch takes field runoff to a nutrient treatment wetland, whose native plants clean the water and reduce erosion. *Courtesy of Emma Dorn, Matt Haken, ISG.*

Saving seagrass

DISCIPLINE Oceanography   **INDUSTRY** Conservation

ORGANIZATION Student   **LOCATION** USA

Many of us are aware of rising sea levels and growing pollution in our oceans, but few know about the threats to one of the most critical ecosystems along our coasts: seagrass beds. The beds provide nutrients for a range of marine and avian wildlife, sequester carbon, and moderate ocean acidification. At the same time, these seagrass beds have suffered from declining density and health because of *Labyrinthula zosterae*, or wasting disease.[2] This story map by a graduate student researcher documents a project that uses drones to provide high-resolution imagery that ArcGIS Drone2Map® can stitch together into a mosaic of images that can help track where and by how much seagrass beds have declined. It's an excellent example of how drone technology can reveal phenomena that would otherwise go unnoticed or, as they say, get lost in the weeds.

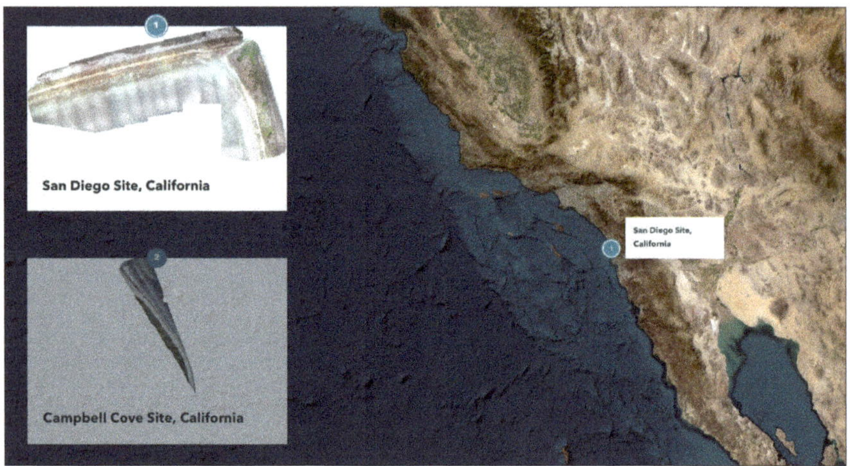

Drone surveys of coastal seagrass beds document their decline resulting from a wasting disease. *Courtesy of Tyler Copeland.*

Global prevalence of languages

DISCIPLINE Computer Science **INDUSTRY** Education

ORGANIZATION Northeastern University **LOCATION** Global

Sometimes, one academic paper can open a whole new field of inquiry. That was the case with the 2013 paper titled "The Twitter of Babel: Mapping World Languages Through Microblogging Platforms," whose authors mined the social media platform Twitter (now called X) over two years and looked at roughly 6.5 million GIS-tagged tweets per day to develop a detailed map of the languages used in more than 100 countries. The results of that research allow users to select a country to see what percentages of what languages are used where, at least among Twitter users. According to the research, a diversity of languages is spoken almost everywhere, even in places long thought to be monolingual. The world is much more complex than many previously thought.

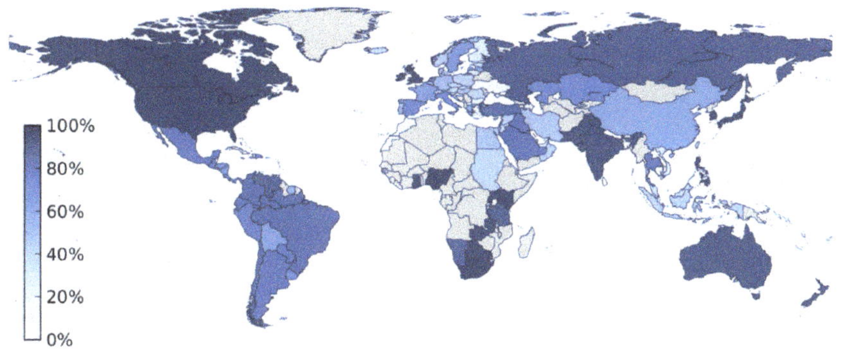

This map shows the languages spoken by Twitter users in each of 100 countries around the world. The color of each country indicates the fraction of users adopting the official language in tweets. *Courtesy of Delia Mocanu, Andrea Baronchelli, Nicola Perra, Bruno Gonçalves, Qian Zhang, Alessandro Vespignani.*

Coastal mangroves

DISCIPLINE Forestry

ORGANIZATION Student

INDUSTRY Conservation

LOCATION Florida, USA

Globally, coastal flooding stems from melting ice caps, but mangroves—salt-tolerant trees and shrubs—can reduce the impact of storm surges, erosion, and rising seas while also filtering pollution, trapping sediments, and providing marine and bird habitat. Mangroves also absorb the atmospheric carbon contributing to climate change. This story map by a student of population medicine maps the loss of mangroves in parts of the planet experiencing sea-level rise and public health impacts from air and water pollution. The story map describes the health impacts of pollution on vulnerable communities, giving a human face to environmental degradation. This ArcGIS StoryMaps Competition winner concludes by showing where mangrove planting along coastal Florida could reduce flooding. As this map suggests, the health of the planet begins with the health of its people.

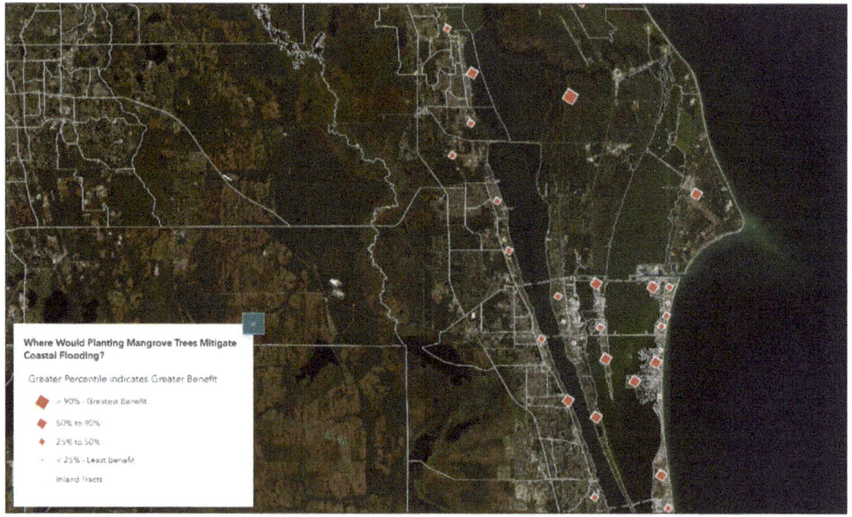

Mangroves are key to improved public health as well as reduced coastal flooding and increased habitats for coastal wildlife. *Courtesy of Vaidehi Patel, 2024.*

Saving sea turtles

DISCIPLINE Biology   **INDUSTRY** Conservation

ORGANIZATION Student   **LOCATION** Spain

The ecosystems that attract marine animals also attract fishing fleets, to the detriment of both. This ArcGIS StoryMaps Competition winner by a biology student looks at how fishing nets trap turtles and prevent them from surfacing to breathe. It also explores the role that sea turtles play in coastal ecosystems and how Spain's Catalan coast has witnessed a high level of turtle deaths due to fishing practices. The solutions include expanding the marine protected areas where sea turtles and other marine species live, restricting fishing access, outlawing more harmful fishing practices, and engaging communities and nonprofits to help distressed animals. The author ends with an action protocol for protecting marine animals, a process that applies not just to the Catalan coast but to every coastal area experiencing conflicts between people and marine life.

The expansion of marine protected areas along the Catalan coast would help reduce the number of sea turtles trapped in fishing nets. *Courtesy of Ona Santisteban.*

Protecting crustaceans

DISCIPLINE Oceanography

ORGANIZATION Student

INDUSTRY Conservation

LOCATION Port Townsend, Washington, USA

Our trash says a lot about us. The paradox of crab pots abandoned by fishing boats is that they reduce the population of the very crustaceans that the operators of these boats depend on for their livelihood. This story map by students who are part of the Sea Dragons, a student-run underwater robotics team, examines the problem of derelict crab pots degrading marine biodiversity and harming the economy of fishing communities such as Port Townsend, Washington. The authors also map the location in the Port Townsend Bay of abandoned pots; an estimated 12,000 pots are abandoned annually, killing nearly 180,000 crabs and other species each year. The narrative describes efforts by community volunteers working with the county and other organizations to recover pots and an initiative to teach residents (including schoolchildren) about the dangers of derelict crab pots in their hometown bay.

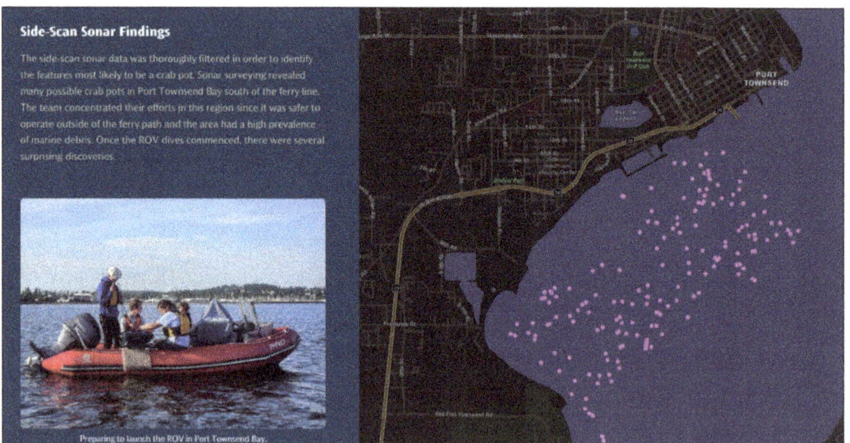

Sonar surveys have identified the derelict crab pots in Port Townsend Bay, which degrade the marine life that residents depend on. *Courtesy of Ella Ashford, Riley Forth.*

Reducing soil erosion

DISCIPLINE Soil Science

ORGANIZATION Student

INDUSTRY Conservation

LOCATION Czechia

Few things degrade soil health more than wind erosion, but GIS allows a level of erosion analysis not possible before. In Czechia, 30 percent of agricultural land is threatened by wind erosion. This story map, by a student at Mendel University in Brno, applies a wind-erosion equation to GIS to visualize the susceptibility of soil to wind erosion in a 52-square-kilometer area in the country's South Moravian Region. The author calculates the impact of seven-meter-wide, 20-meter-high windbreaks in strategic locations and shows the benefits of an investment in these breaks. This study recognizes that those benefits take time to manifest, although the costs of installing them are incurred immediately, meaning that long-term thinking and capital are required to protect soils. Once the soil is gone, it will take far longer to replace it, which makes any effort at reducing erosion the most cost-effective effort of all.

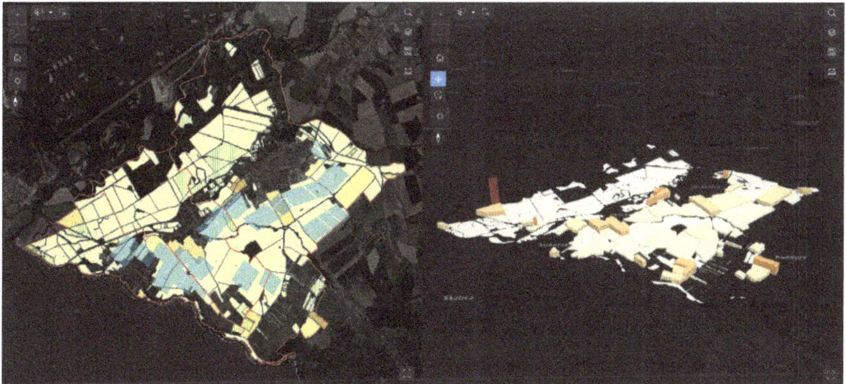

Windbreaks offer one of the most effective ways to reduce soil erosion. These maps show those benefits in useful ways. *Courtesy of Petr Zalesak.*

Notes

1. The map and chart data comes from the 2022 County Health Rankings dataset.
2. Dr. Bo Yang, byang85@ucsc.edu: https://agupubs.onlinelibrary.wiley.com/doi/full/10.1029/2022GL101985.

References

Access to health care: https://storymaps.arcgis.com/stories/94315ac008b947f9b1cafa930b01832f

African American golfing: https://storymaps.arcgis.com/stories/98e108146e954e84a36d57f6afd6b362

American jazz clubs: https://storymaps.arcgis.com/stories/9668519a946646bcaa85aa977afff0fa

Bike access in New York City: https://storymaps.arcgis.com/stories/0c76de4893074bd8825252386fb91163

Coastal mangroves: https://storymaps.arcgis.com/collections/40cf74f19e884724a70f4e1bb4e4adfa?item=9

Disappearing farmland in the West Bank: https://storymaps.arcgis.com/stories/59fe0ecd40c1455cb8f9b9f380c48207

Enslaved peoples' migration: https://www.bunkhistory.org/resources/southern-history-the-migrations-of-the-american-south-1790-2020

Esri: esri.com

Femicide in Türkiye: https://storymaps.arcgis.com/stories/c51f1bdcf12240ae8cc5c3e541ed06bd

Geolocating Atlanta's rap: https://storymaps.arcgis.com/stories/938c1fd29e0045c8a3620196af3f36e3

Global prevalence of languages: https://journals.plos.org/plosone/article?id=10.1371/journal.pone.0061981#authcontrib

Mapping Dante's *Inferno*: https://storymaps.arcgis.com/stories/ad2a09720b75435b922396307e2d6004

Mapping James Joyce's *Ulysses*: https://www.michaelgroden.com/notes/map.html

Mapping medicine's consolidation: https://storymaps.arcgis.com/stories/636f8a9b59b94d81bdd09b80a9e531e6

Mapping mental health: https://storymaps.arcgis.com/stories/f991326818c24aa1852335d1ef45f23b

Mapping temporary memorials: https://storymaps.arcgis.com/stories/7ccaa25885fa4c94992b5082ffda7e9a

Margarette Nevalainen: https://www.linkedin.com/in/margarette-nevalainen/

Mortality in medieval Europe: https://storymaps.arcgis.com/stories/1bcbbee6d0b64f1eb8a499ad096e176c

Nightlife in old Shanghai: https://www.esri.com/about/newsroom/app/uploads/2019/11/teachingSM.pdf

Northeastern University: https://www.northeastern.edu

Ownership of Indigenous land: https://storymaps.arcgis.com/stories/a7ea2f7b0b184d838772360f9833fd36

Painting with GIS: https://www.esri.com/arcgis-blog/products/arcgis-pro/mapping/painting-in-pro

Pediatric malaria in Nigeria: https://storymaps.arcgis.com/stories
    /a296600c350f4684b3087d1e38194c02
Preserving Native American sites: https://storymaps.arcgis.com/stories
    /980f010ccbe14ebe8635e99d8d89a917
Protecting crustaceans: https://storymaps.arcgis.com/stories
    /1befb7cae32f49e89d8595d8ae884d38
Reducing soil erosion: https://storymaps.arcgis.com/stories
    /c3bdb3174746476780d4547e5eea4ebd
Saving sea turtles: https://storymaps.arcgis.com/collections
    /40cf74f19e884724a70f4e1bb4e4adfa?item=11
Saving seagrass: https://storymaps.arcgis.com/stories
    /e554d25d27264872b3a02daa0bc91ecb
Shared feelings about places: https://storymaps.arcgis.com/stories
    /7baa6133dd6b49bc80971a7e7abf2353
Spanish silver trade: https://www.esri.com/about/newsroom/app/uploads/2019/11
    /teachingSM.pdf
The culture-nature divide: https://storymaps.arcgis.com/stories
    /ccff29d28b6d472086885d26f1bc3f28
Tracking bird strikes: https://storymaps.arcgis.com/stories
    /36fb13e177a345a7adfe14e02bac52b2
University of Minnesota: https://umn.edu
University of Richmond: https://www.richmond.edu
Water quality in farm fields: https://storymaps.arcgis.com/collections
    /40cf74f19e884724a70f4e1bb4e4adfa?item=13

Introduction

The facilities staff at most colleges and universities, in the more developed countries, have long used GIS to help them track and manage the many physical assets of most higher education institutions. Despite that history, however, there remains much to learn. This chapter includes various examples, explained in greater detail after this introduction, of how a range of colleges and universities have used geospatial tools—apps, story maps, dashboards, and the like—to gain more efficiency, accuracy, and agency in their work.

Higher education institutions handle their facilities in a variety of ways. Almost all have their own in-house maintenance crews, and most have in-house campus planners and facilities staff, although many also use outside vendors and consultants, especially for larger projects. Several design, engineering, and planning firms also specialize in higher education work, evident in some examples in this chapter. Campuses are also public spaces, even when the institution is private, and designing them to accommodate visitors as well as faculty, staff, and students can be a challenge, aided in all cases by GIS and the access to the information that it provides.

The changing context of higher education has also brought a rethinking of how much and what kind of facilities colleges and universities will need in the future. The growth in online or distance learning, the availability of remote library or laboratory resources, and the increase in active learning or flipped classrooms have forced institutions to renovate antiquated buildings and repurpose a lot of underused space. Here, too, GIS has proven invaluable as an interior mapping and data-tracking tool to assess space use and align space needs with the available facilities. It's all about facilitation.

Campus access

One of the most visible uses of GIS on campus involves helping not just students and staff but also visitors find their way around places that, because of their often parklike settings and pedestrian-oriented spaces, can seem disorienting to those unfamiliar with them. The emergence of digital maps has greatly increased people's ability to access campus facilities. Enhanced by great cartography, such as the navigation system at the University of Texas at Austin, these GIS tools can help campus visitors understand where they are and where they want to go.

Maps have also helped people understand other barriers that can make traversing a campus so challenging. The degree of slope and the amount of grade change on hilly campuses can present one type of obstacle, which the University of California (UC), Berkeley, has addressed with its accessibility map, showing where the low-slope paths and places can be found. Locating specific facilities for certain groups offers another challenge, which the University of Michigan has addressed with its crowd-sourced map of gender-neutral spaces on campus. And finding places of respite and renewal on campuses typically buzzing with activity is also something that benefits from GIS, as discussed in the section about the green space map that North Carolina State University has developed with student mental health in mind.

Getting to and from campus can be as challenging as navigating it once there. Georgia Tech's transportation management system shows visitors when and where to park, which is especially useful during periods of peak demand. And finding the right door to enter or the right room to access makes wayfinding, both indoors and out, difficult for any first-time visitor or first-year student. The development of mapping tools that show users how to get where they're going has made that much easier, as the examples of Boston College and the University of Applied Sciences in Dresden, Germany, demonstrate.

Campus assets

A less visible but no less important use of GIS for campus facilities involves the mapping of an institution's assets. This can include mapping the flow of water in a stormwater system, as the Mississippi Watershed Management Organization has done in partnership with the University of Minnesota, or identifying the location of key underground utilities in a digital twin of the campus, as the University of Kentucky has done, or creating an atlas of every asset, above and below ground according to a campus grid, as UC Santa Barbara has done.

But a campus involves not only its pipes and tunnels. Its key assets are the buildings and spaces where teaching and learning occur; mapping indoor spaces has become one of the most significant new applications of GIS on campus. These applications range from (a) the integration of building data into a comprehensive map of the campus facilities, such as Michigan State University's mobile asset map, to (b) mapping actual space use, such as The Ohio State University's interactive space-use app, to (c) tracking a campus's space-use efficiency, such as the space-time analysis of the State University of New York (SUNY) at Cortland's facilities, to (d) a dashboard that can inform everyone on campus about the use of rooms, such as Arizona State

University's space management system. The real value of these asset-management tools lies not just in the day-to-day operation of a campus but also in extraordinary situations, underscored by the University of South Florida's priority map of facilities to protect when threatened by a hurricane.

Campus quality of life

A third area of GIS use on campus involves quality-of-life issues. These can entail mapping and tracking the health of the campus, be it the physical and mental health of the campus community or the environmental health and social impact of campus facilities, as the University of California, Los Angeles (UCLA) has achieved with its health campuses. Food insecurity remains another quality-of-life challenge on many campuses, leading institutions such as the University of Redlands to set up a farm on campus to address the needs of nearby, underserved communities for food and shade. Mapping a campus's tree canopy and informing the community about the university's public art, as the University of Minnesota has done, offers yet another way to see the campus as an agent for the good of communities.

The sheer density of people on campus can lead to some negative outcomes as well. One such example is the garbage that people leave on the grounds, which St. Mary's University in Winona, Minnesota, has mapped to know where to install more trash receptacles on campus. Another one is theft. The University of Massachusetts at Lowell has developed a web-based interface that lets the university identify where theft has occurred and what has been stolen to know what needs replacing. All of this shows the remarkable impact that GIS has had on the operations of campuses and on people's experience of them.

Campus planning

GIS began as a planning tool, and it continues to play a vitally important role in campus planning. Geospatial tools have let campus planners see the possibilities—and achieve the efficiencies—of shared resources among different institutions, as the Auraria Higher Education Center in Denver has done with three institutions sharing support facilities. Geospatial technology has also allowed universities to incorporate qualitative and quantitative information in envisioning their future, as the University of New Hampshire has done by embracing Indigenous ways of knowing and understanding the land in its campus plan. The amount and scope of the data that GIS can leverage also makes it an excellent regional-planning tool, which campuses such as Dartmouth College in New Hampshire have used to understand the connections and opportunities that exist among its various properties extending over a broad landscape.

Geospatial technology also allows campus planners to abstract information about a place and reframe it in new ways. Emory University's Framework Plan, for example, approaches the campus not just as a physical place but a conceptual one, in which the connections among disciplines and the interactions of units are important to achieving its "One Emory" goal. Digital dashboards, such as George Mason University's, also make data available to a greater diversity of people and make the institution more transparent about its operations and decision-making.

GIS continues to play a role in the physical design of campuses, making that undertaking more grounded in data and more encompassing. GIS-based maps can highlight the physical connections within and among campuses and with the surrounding community, as the campus plan for the University of Minnesota suggests. Data-based maps can also inform an institution about how and where it might expand in the future, as happened at the University of the West Indies. And finally, GIS can help institutions determine where an entirely new campus might be built, as in the case of Sacramento State's Placer Center, and what it might look like once built, as in the case of the Joi Institute in Mumbai, India.

The spatial turn that has characterized so many disciplines in higher education has equally characterized the operations and planning of campuses. GIS has become the common underpinning of the intellectual and physical aspects of colleges and universities, and because of that—and contrary to what some of the critics of higher education might say—these institutions are more efficient and effective, accessible and responsive, and flexible and adaptive than ever before, as the following examples suggest.

Campus access examples

Mental health

DISCIPLINE Public Health   **INDUSTRY** Facilities

ORGANIZATION NC State University   **LOCATION** Raleigh, North Carolina, USA

Growing numbers of college students are facing mental health challenges, and the physical campus itself can help higher education institutions address this need. North Carolina State University has mapped its green spaces, not for operations or facilities management purposes but as a guide for students seeking the stress relief that can come from being in natural settings. The map shows all the campus green spaces, with filters that allow students to sort the data according to their need to relax, socialize, study and teach, eat, walk and bike, play sports, or explore. The map's sidebar has clickable information about each space, focusing on what the space has to offer students looking for places to reduce the pressures they feel. Each sidebar has suggestions for other places students might also want to explore as well as an image of each space, bringing the map to life. We tend to think of campus facilities in terms of built infrastructure, but this mapping effort shows how much campuses can also facilitate better mental health for the people who work or study there.

Indoor wayfinding

DISCIPLINE Campus Operations

ORGANIZATION University of Applied Sciences in Dresden

INDUSTRY Facilities

LOCATION Dresden, Germany

Finding one's way in many campus buildings can be a challenge even for spatially oriented people, in part because university structures often change over time or because they have many spaces off-limits to students or visitors. An undergraduate student at the University of Applied Sciences in Dresden, Germany, developed a web app and mobile app, using ArcGIS Indoors, that enables people to find the best routes in and around buildings and to get to their destinations, with clear instructions on where to make turns and what distances lie ahead. That an undergraduate student could create such applications shows not only how easily configurable a platform like Indoors can be but also how much students recognize the need for easily used navigation tools, something that faculty and staff familiar with campus buildings may not appreciate. The first step in learning requires that people can get to the place in which the learning will occur.

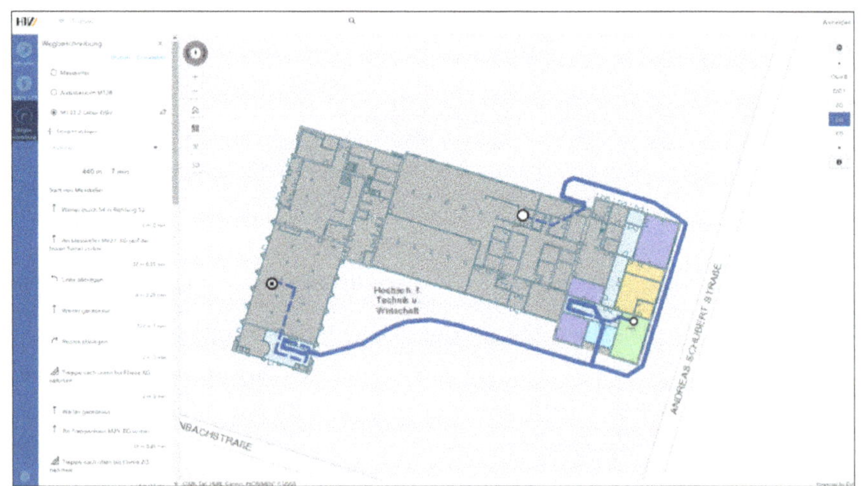

This image shows how the tool can direct a person to a destination along the most direct and efficient path, showing the building's floor plan and which spaces are accessible. *Courtesy of Ricardo Roch, University of Applied Sciences, Dresden.*

Accessibility

DISCIPLINE Campus Operations
INDUSTRY Facilities
ORGANIZATION University of California, Berkeley
LOCATION Berkeley, California, USA

Many campuses occupy sites with significant grade changes, making movement across them a challenge for people who have mobility limitations. UC Berkeley's campus has a 250-foot grade change across its Campus Park. So, working with the planning firm Sasaki, the university has laid out the paths and places on campus that have slopes of 5 percent or less. To develop these accessibility maps, the design team toured the campus with stakeholders and leaders to understand their lived experiences, used virtual reality to help people envision design options, and developed tactile models and braille maps to engage people with visual impairments. They also put on a public event with national accessibility leaders to share information and ideas related to human-centered design and the history of disability rights. Accessibility requires care in the paths and places we create for people, just as the process of making campuses accessible is itself a journey down a path, to a place where everyone is accommodated.

Inclusivity

DISCIPLINE Campus Operations
INDUSTRY Facilities
ORGANIZATION University of Michigan
LOCATION Ann Arbor, Michigan, USA

Volunteered geographic information (VGI) offers a bottom-up approach to mapping that can save organizations and institutions staff time and produce maps that are more meaningful and sometimes more accurate than the official maps of a place. Staff at the University of Michigan created a crowdsourced map of the campus that allowed members of the university community to locate spaces that matter to different users, such as gender-neutral restrooms, lactation rooms, and reflection rooms. They developed a mobile-friendly web map using ArcGIS tools and added filters so that people can search for only what they need to find. To make sure the system is easy to use and update, the team also created a database that stores the spaces and their attributes in ArcGIS Online and configured ArcGIS Collector to allow people to crowdsource additional information. By letting people across campus contribute to this mapping project, the university has helped make VGI stand for something else: a Very Good Investment.

Navigation

DISCIPLINE Campus Operations

ORGANIZATION University of Texas at Austin

INDUSTRY Facilities

LOCATION Austin, Texas, USA

Large campuses can pose a challenge to incoming students and visitors alike, and maps can help people navigate spaces and buildings that might otherwise be disorienting. The University of Texas at Austin, as part of its ongoing integration of GIS into its facilities and operations, created a clear and convenient GIS map of its campus, with a set of search buttons and drop-down menus that allow people to find not only where they want to go but also to locate important services, from ATMs and health services to bike-share stations and dining facilities. Large campuses are like miniature cities, and geospatial tools can make them less daunting and more accessible.

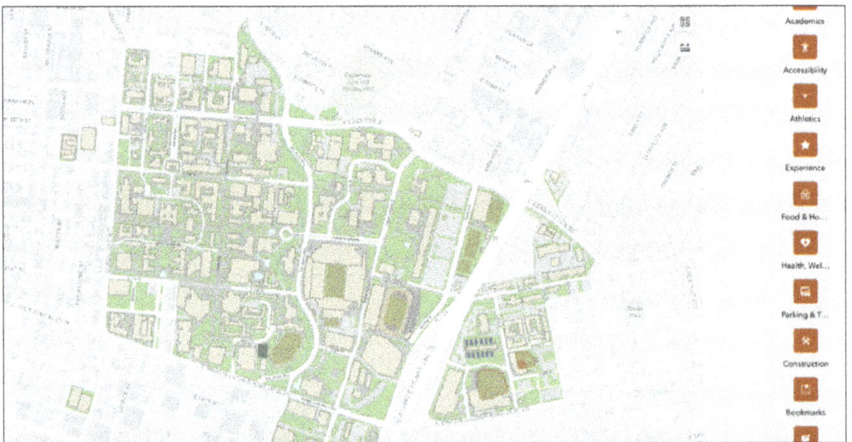

The University of Texas at Austin developed an interactive map that helps people navigate the campus according to a variety of search buttons, leading to drop-down menus, on one side.

Copyright © The University of Texas at Austin.

Wayfinding

DISCIPLINE Campus Operations **INDUSTRY** Facilities

ORGANIZATION Boston College **LOCATION** Boston, Massachusetts, USA

The Americans with Disabilities Act (ADA) has led campuses across the United States to accommodate people with special needs, although locating accessible entrances and access points remains a challenge for some users. Boston College has developed an ArcGIS Enterprise application that stores and enables better control of all facilities data and makes it easier to share accurate information—in this case about accessibility routes and entrances to all its buildings and grounds. Each data point identifies not just the location but also the character of the entrance—whether an entrance is power-operated or not, for example—and any other comments about it that might be helpful. The map also identifies elevators inside buildings. Although the ADA has made campuses more accessible, GIS can make them more legible for people of all abilities.

Parking

DISCIPLINE Campus Operations **INDUSTRY** Facilities

ORGANIZATION Georgia Tech **LOCATION** Atlanta, Georgia, USA

Having enough parking in the right locations for staff, students, and visitors coming to a campus remains a problem, regardless of the size of the institution. Georgia Tech has instituted a transportation management system that lets it accommodate peaks in parking demand without major increases in the amount of parking infrastructure. By mapping the parking demand by the date and time of day, the university saw opportunities that might not have otherwise appeared, such as creating pay-as-you-go parking close to campus, producing better mobility and pedestrian experiences for staff and faculty to incentivize them to park in more distant locations, and guiding people to the best parking locations using digital wayfinding signage and digital mobile guidance systems. Transportation management initiatives like this can not only reduce the demand for more parking facilities but also enhance the experience of campus visitors as well as the people who work or learn there.

Campus asset examples

Digital twins

DISCIPLINE Campus Operations **INDUSTRY** Facilities

ORGANIZATION University of Kentucky **LOCATION** Lexington, Kentucky, USA

Digital twins allow campus facilities staff to understand the relationships among the many systems and structures that make up university campuses. The University of Kentucky has developed a virtual replica of its campus that includes precise information about its infrastructure, facilities, utilities, and natural surroundings, which in turn enables a more efficient and effective operation of the campus. The digital twin allows facility managers to search for a variety of systems, ranging from built infrastructure and physical structures to telecommunications locations, such as communications equipment. The digital twin also allows the university staff to run future scenarios to test the resilience of its systems and respond to incidents on campus with unprecedented situational awareness. Creating a digital twin can take time and effort, but once established, it can have a lot of benefits. As with human twins, digital ones present a mirror image that helps organizations see themselves more clearly.

Asset mapping

DISCIPLINE Campus Operations **INDUSTRY** Facilities

ORGANIZATION Michigan State University **LOCATION** East Lansing, Michigan, USA

Campuses contain thousands of physical assets—buildings, roads, utilities, trees, and so on—and keeping track of the location and condition of those assets can seem overwhelming, especially for large campuses. Michigan State University has undertaken an asset mapping initiative identifying tens of thousands of assets that can be viewed not only on a desktop computer but also on a mobile device so that staff can view them in the field. ArcGIS Indoors offers an off-line mobile app that shows not only the assets anywhere on campus but also directions to a particular room and where to park. It has allowed employees, when on location, to turn on and off other types of information or unrelated data about a place to focus on their work. The ability to have all the assets in an already mapped location has been a game changer for Michigan State's staff, increasing efficiency and avoiding unnecessary problems.

Water

DISCIPLINE Environmental Science

ORGANIZATION Mississippi Watershed Management Organization

INDUSTRY Water

LOCATION Minneapolis, Minnesota, USA

Campuses, especially in climates with a lot of precipitation, shed a lot of stormwater into the larger bodies of water around them, and mapping that process can reveal the impact of stormwater decisions on watersheds. The University of Minnesota developed a stormwater tracking app for the Mississippi Watershed Management Organization (MWMO) that allows users to pick any location on or near campus within the MWMO region and see the route that the stormwater takes to get to the Mississippi River, how long that runoff takes, and what condition the water will be in when it drains into the river. The power of a mapping app like this is that it makes what is often invisible to people and organizations more visible and tangible, which is the first step in changing behavior and prompting investment.

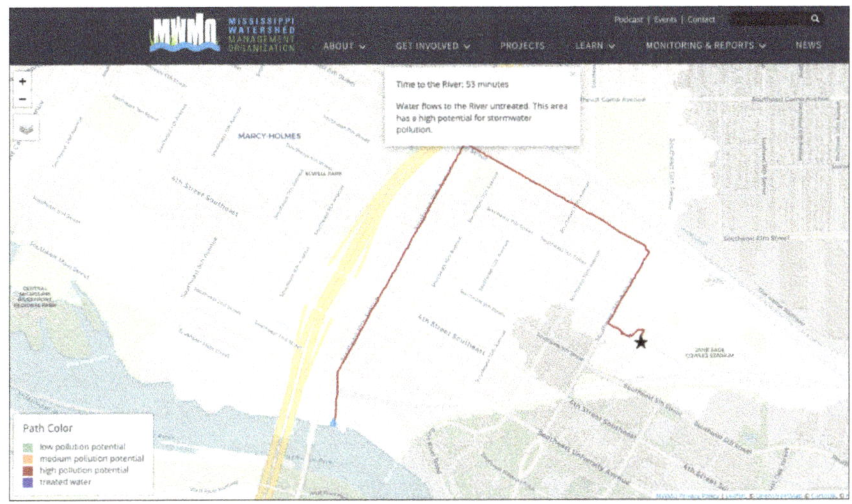

The University of Minnesota developed this app for the Mississippi Watershed Management Organization to help individuals and organizations track the impact of stormwater pollution from a given location. *Courtesy of the University of Minnesota.*

Utilities

DISCIPLINE Engineering

ORGANIZATION UC Santa Barbara

INDUSTRY Facilities

LOCATION Santa Barbara, California, USA

Campuses often have complex belowground utilities serving numerous types of buildings found at most universities. Keeping track of the different types, ages, and conditions of all those pipes and conduits can be a headache. The University of California, Santa Barbara, had the special challenge of occupying a former military base, with utilities from that era still in place. To keep track of the campus's electric, natural gas, domestic water, reclaimed water, chilled water, seawater, sewer, and storm lines, the university created an atlas of all its holdings, with a grid of more than 200 units, each of which has an aerial image and a full-color and line work map of every utility in that square. This not only makes the complex utility network more manageable but also eases the updating of the atlas each month. It also enables the use of ArcGIS Field Maps so that workers on site can update the atlas as changes occur. It's a solution that would have made the former occupants of the site—the military—proud.

The utility atlas of the UC Santa Barbara campus makes belowground utilities easier to locate, track, upgrade, and maintain. *Courtesy of Paul Bartsch, GIS Programmer, University of California, Santa Barbara.*

Emergency management

DISCIPLINE Emergency Management   **INDUSTRY** Facilities

ORGANIZATION University of South Florida   **LOCATION** Tampa, Florida, USA

Some campuses occupy locations prone to various natural hazards, from wildfires to floods. The University of South Florida has several campuses in hurricane-prone locations, and its facilities staff have used geospatial tools to track the paths of hurricanes affecting their campus and keep the campus community informed of hazard response. This has led to a strong partnership between the university and local emergency management staff, sharing data that has enabled the area's emergency managers to determine inland flood risks, predict storm-surge impacts, and model the effects of storms before they happen. Geospatial tools have also let the university pull data from various services, from satellite imagery to traffic congestion along evacuation routes, and enabled the facilities staff to prioritize the buildings that have the most vulnerable research facilities to ensure that they remain in operation. South Florida has taken campus safety and weather mapping to a whole new level.

Space use

DISCIPLINE Campus Operations   **INDUSTRY** Facilities

ORGANIZATION Ohio State University   **LOCATION** Columbus, Ohio, USA

Few topics trigger the territoriality of faculty and staff more than space allocation and use, even though physical space has become less central to the educational enterprise as remote teaching and learning have become more common and research and scholarship have become more virtual. The Ohio State University, one of the largest campuses in the United States, has developed an app that allows its facilities staff to track space use, including who occupies what space, what space gets allocated to what budget, and how much space gets used for what purpose. Heavily used by the university's planners, this space-mapping tool also allows faculty and staff to understand their own space use and to see how it compares with that in other departments. Such applications reveal the power of bringing geospatial tools into interior environments. In an era in which digital tools have begun to transform teaching and learning, these indoor applications will become ever more important as we track the impact of the digital educational environment on physical space.

Space efficiency

DISCIPLINE Campus Operations   **INDUSTRY** Facilities

ORGANIZATION SUNY Cortland   **LOCATION** Cortland, New York, USA

Even as most colleges and universities track their space use, many may not know how that use compares with other institutions or even with their own space use over time, which contributes to the paradox of ever greater demand for space on campus, even as a lot of space goes underused and as remote or off-campus learning becomes more common. SUNY Cortland has used ArcGIS Indoors and developed an ArcGIS Experience Builder app to aid everyone's understanding of actual space use over time. The space-management tool lets department administrators propose and review space and room-numbering changes and track vacant or underused spaces that may be charged to a particular department. The app also allows people to look for space that might be available when they need a space to meet or work. All this increases the availability of space without having to add square footage. As the rest of the work world becomes accustomed to coworking in shared spaces, the academic world will need to follow, and geospatial tools can help academic staff find the spaces they need—and those that they no longer do.

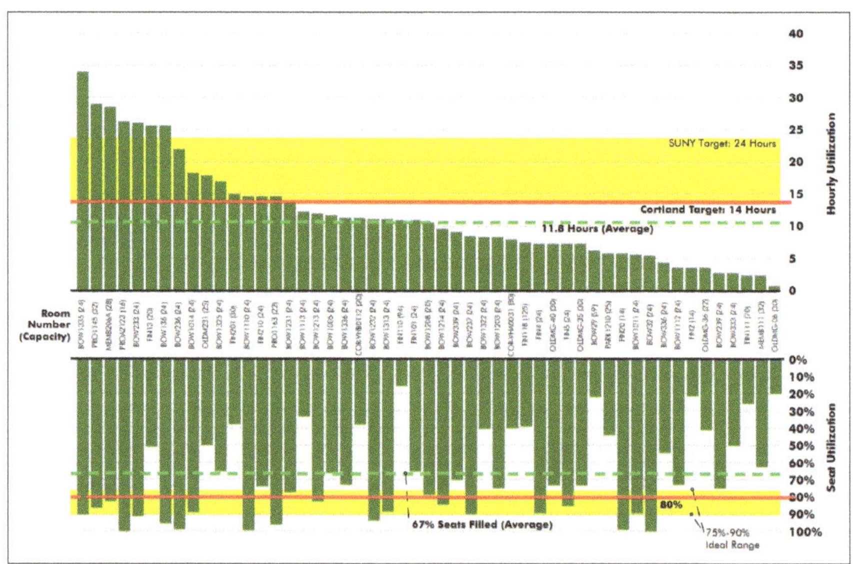

By mapping space utilization and graphing it against a target, SUNY Cortland now has the tools it needs to track underutilized spaces and craft better ways to use the space that it **already has.** *Courtesy of the SUNY Cortland, JMZ Architects and Planners.*

Space management

DISCIPLINE Campus Operations   **INDUSTRY** Facilities

ORGANIZATION Arizona State University   **LOCATION** Tempe, Arizona, USA

The dynamics of space use on a campus make managing that space a constant struggle. Arizona State University has addressed that challenge with the development of tools to make space management easier, with its Space Management dashboard. By mapping every building, Arizona State has given academic administrators and facilities and public safety staff alike the ability to see who is doing what and where. The university has also built an ArcGIS Experience Builder web app that lets users interact with the tool to look at space use in various ways— for example, through using the national facilities inventory code (or PEFI). Custom widgets in the app also highlight data for a particular user or summarize space assignment statistics for a particular floor in a building. GIS becomes like a library catalog for campus space, enabling different people to find the information they need when they need it.

Campus quality of life examples

Health

DISCIPLINE Public Health   **INDUSTRY** Health and Human Services

ORGANIZATION UCLA Health   **LOCATION** Santa Monica, California, USA

Health on campus has become a priority in higher education as data shows worrying trends related to the mental health of students, the environmental health of buildings, or the health and wellness of staff. The UCLA Health system has created a series of dashboards that track its progress on achieving better health in several areas: climate resilience, energy conservation, green buildings, green operating rooms, nutrition and wellness, sustainable procurement, sustainable transportation, waste reduction, and water conservation. As a member of the health-care sector's sustainability initiatives, called Practice Greenhealth, UCLA Health not only reports annually on its progress but also makes these dashboards available to the community to track the impact of the system's sustainability work. Although many universities participate in programs like this, UCLA's efforts have the noteworthy quality of bringing all this information (much of it geospatially based) together in one place, in a series of well-designed and easily understood dashboards. It's a healthy model.

Tree canopy

DISCIPLINE Forestry

ORGANIZATION University of Minnesota

INDUSTRY Education

LOCATION Minneapolis, Minnesota, USA

Every campus has its distinctive plant communities, which enhance the campus experience but are rarely described in campus literature or highlighted on campus tours. This campus tree tour of the University of Minnesota shows how mapping the trees on campus can inform the public and the campus community about the diversity of plant life at the university and engage people in the rich history and rigorous science of forestry. This effort demonstrates how mapping can be a way to link the facilities at a university with the knowledge of its faculty and to make the campus itself part of the learning experience of students. The project also exemplifies the way in which compelling images, clear writing, and informative maps can complement one another to tell stories that have visual and intellectual appeal. Every campus should conduct such a tree tour.

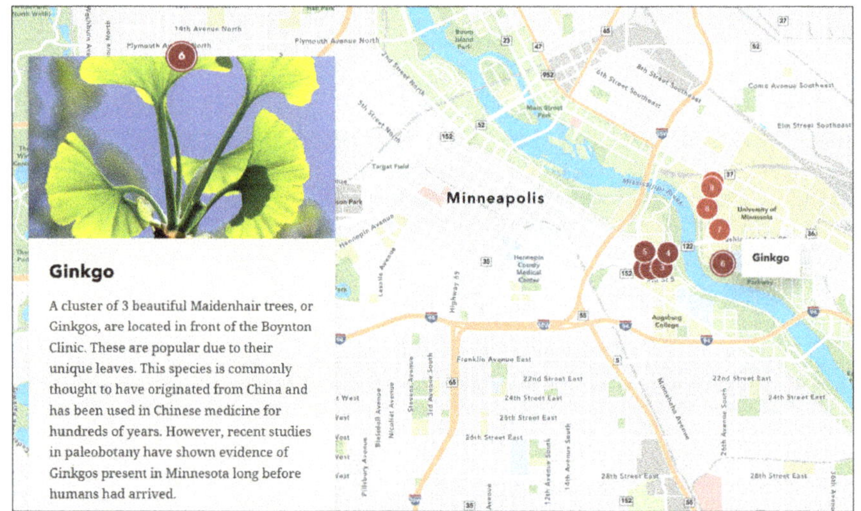

The University of Minnesota's tree map highlights the diverse tree species on campus, provides information about each species, and locates examples on the campus map. *Courtesy of the University of Minnesota.*

Theft

DISCIPLINE Campus Operations   **INDUSTRY** Public Safety

ORGANIZATION LandTech   **LOCATION** Lowell, Massachusetts, USA

Although permanent assets on a campus are hard enough to track and manage, the electronic assets on campus pose an even greater challenge because they're easier to steal and harder to notice if they're not present. The University of Massachusetts at Lowell worked with the consulting firm LandTech to develop an ArcGIS Indoors app that allows universities to track assets and support facility management, planning and scheduling, operations, security, and wayfinding. For example, the ability of the system to note that a piece of equipment has disappeared and show where it was located helps facilities staff, who must replace the technology, and campus security. Maps really help when, as Joni Mitchell put it, "You don't know what you've got till it's gone."

Food

DISCIPLINE Agriculture   **INDUSTRY** Education

ORGANIZATION University of Redlands   **LOCATION** Redlands, California, USA

Redlands, California, like many cities, has a more affluent section, with tree-lined streets and ample green space, and a more impoverished one, with fewer trees, less shade, and little green space, in a food desert. To address that discrepancy, the University of Redlands established a sustainable farm in partnership with California's Climate Action Corps Fellows to bring together staff and students from the university and members of the Redlands community to grow food, compost waste, and create a tree nursery. The farm teaches farming skills and generates produce for the community. The founders of the farm have used geospatial tools to identify the parts of the city that have food and tree deserts as the basis for enlisting community members and distributing produce. The group has also worked with the Native American community to grow and use medicinal plants and with faculty at the university to monitor butterflies and assess drought-tolerant plants. A campus is not just a facility—it can also be a farm.

Public art

DISCIPLINE Arts	**INDUSTRY** Museums
ORGANIZATION University of Minnesota	**LOCATION** Falcon Heights, Minnesota, USA

In the United States, 27 states have percent-for-arts programs, and many are like the one in Minnesota, which requires the allocation of up to 1 percent of the budget for construction or renovation of $500,000 or more be spent on public art. The Weisman Art Museum at the University of Minnesota has published a story map that locates all the public art projects on campus, with written and audio descriptions of the artwork and the artists and a photograph of the art itself. This site turns the campus into an outdoor exhibition space and provides explanations of the artists' intentions and interpretations of the work for those who walk by these works every day and might wonder about their meaning. The relation of the art to the activities conducted in the adjacent structures also helps reveal the purposes of campus buildings to passersby. Campuses need curation as much as the museums they often house.

Minnesota has a requirement for public art accompanying all new public campus buildings, and the Weisman Art Museum at the University of Minnesota has made a story map locating and describing all the campus art installations. *Courtesy of Weisman Art Museum.*

Trash

DISCIPLINE Campus Operations

ORGANIZATION St. Mary's University

INDUSTRY Education

LOCATION Winona, Minnesota, USA

Many universities occupy sites with a lot of natural resources on or close to campus, and preventing trash from polluting those landscapes, visually as well as environmentally, remains an ongoing effort of staff and students alike. St. Mary's University of Minnesota shows how mapping can help in that work. Capitalizing on its location in the low hills and valleys along the Mississippi River, the university seeks to be an "eco-campus," with maps showing the diverse flora and fauna on campus and what steps the university is taking to protect them. The university has also developed a heat map showing where trash tends to collect around heavily used areas of campus and overlaid that heat map with the location of trash containers, identifying where more of the latter would improve the cleanliness of the campus. Finally, the university has made this map publicly available to raise awareness and change the behavior of those responsible for the trash. Cleaning up a campus requires that some people clean up their act.

St. Mary's University of Minnesota mapped areas of trash accumulation and overlaid that heat map with a map of trash can locations to determine where more of the latter were needed.

Courtesy of St. Mary's University of Minnesota.

Campus planning examples

Dashboards

DISCIPLINE Campus Operations **INDUSTRY** Education

ORGANIZATION George Mason University **LOCATION** Fairfax, Virgina, USA

Dashboards have transformed people's ability to access data about a place, and George Mason University's campus plan shows how data can inform decisions about what and where to act. The university, with three distinct campuses, lacked a strategic master plan, so it commissioned a diverse team to develop such a plan by administering a participatory process, with roughly 5,000 people taking part online in interactive mapping exercises. The team then used machine learning algorithms to sort out positive and negative sentiments and analyzed existing space use, program connections, future demographics and enrollments, potential future space needs, financial realities, and physical structures. A dashboard of the final plan makes all that information available to surrounding communities, whose members have the most at stake in making that plan a reality. Such dashboards are a tool that every campus should use.

Shared facilities

DISCIPLINE Campus Operations **INDUSTRY** Facilities

ORGANIZATION Auraria Higher Education Center **LOCATION** Denver, Colorado, USA

Higher education institutions face increasing pressures to be more efficient and collaborative in their use of resources and more distinctive and competitive in their recruitment of students. Three Colorado institutions—the Community College of Denver, Metropolitan State University of Denver, and the University of Colorado Denver—have taken the unusual step of sharing one campus to reduce costs and increase efficiency while retaining their different histories and missions. With Sasaki and U3 Advisors as their consultants, the Auraria Higher Education Center evolved out of an extensive engagement process with all three institutions and their various stakeholders, arriving at a "learning loop" plan where the shared services and programs occupy the center of the campus, with a "front door" to each of the institutions that distinguishes their institutional neighborhoods and reinforces their separate identities. Sharing is never easy, but when done well—as with these institutions—it can make higher education more resourceful and resilient.

Visions

DISCIPLINE Urban Planning	**INDUSTRY** Facilities
ORGANIZATION The Society for College and University Planning	**LOCATION** California, USA

The eastern half of the United States is full of small college towns, reflecting a tradition dating from the nation's founding of colleges serving as magnets to attract new businesses and residents. The plan of a new campus for Sacramento State Placer Center continues that tradition at a scale much larger than in the past. The owner of 2,200 acres in Southern Placer County in California donated 301 acres at the center of an anticipated new town for a new campus for Sacramento State University. The urban design team, led by Sasaki, engaged in a 15-month process, involving multiple stakeholders and advisory groups, to develop a campus plan that respected the rolling grasslands site—a plan that would ultimately serve 25,000 students. The academic plan has six thematic clusters focused on regional workforce needs—resource management, entertainment, wellness, making, practical learning, and general education offerings—and will have facilities that also benefit the community. The plan is to include a hotel, performing arts center, and fire station. Campuses not only serve cities—they can spur the creation of them as well.

The Sacramento State Placer Center will occupy more than 300 acres at the center of a 2,200-acre, mixed-use new town anticipated on land in Southern Placer County, California. *Courtesy of Sacramento State University, Sasaki.*

Frameworks

DISCIPLINE Campus Operations

INDUSTRY Facilities

ORGANIZATION The Society for College and University Planning

LOCATION Atlanta, Georgia, USA

Institutions of higher education have proven to be resilient over the centuries because they provide a framework within which many disciplines, perspectives, and people can thrive. Emory University's Framework Plan exemplifies that idea. Rather than pursue a detailed master plan, the university sought to cultivate a culture of planning, convening conversations among its units and looking for overlaps and common themes that helped set priorities and identify places where different units could integrate and share space. The plan also focused on the student experience by increasing undergraduate and graduate student housing, providing space for diverse cultural affinity groups, and addressing students' wellness needs with meditation, mental health, and recreation spaces. This reflected a goal of creating "One Emory," capturing the strength not just of that university but higher education more broadly, in which campuses are both one thing and many things at the same time.

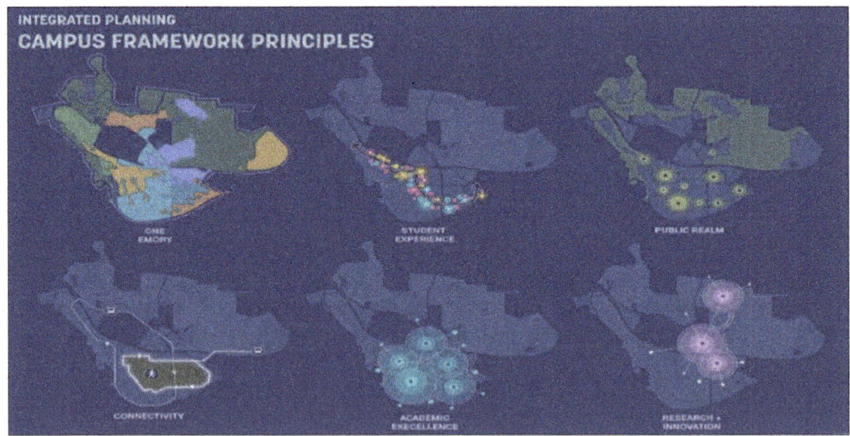

In its goal to achieve "One Emory," the university developed a framework plan that identifies synergies and strengthens relationships among the campus's diverse units. *Courtesy of Emory University, Sasaki.*

Expansions

DISCIPLINE Campus Operations
ORGANIZATION University of the West Indies

INDUSTRY Facilities
LOCATION Trinidad and Tobago

As the global demand for access to higher education grows, especially in developing nations, so too must their universities expand, as is the case for the University of the West Indies, serving the Organisation of Eastern Caribbean States, a nine-member group of islands in the Eastern Caribbean. A 12-month planning process, led by Sasaki, included not just university stakeholders but also community members and government representatives. From that engagement came many ideas, such as having academic buildings usable by small businesses and entrepreneurs and having support facilities, such as the library, student center, and recreation and wellness center, accessible by the larger community. Responding to the tropical climate, the planners envisioned a meandering pedestrian corridor, connecting the expanded campus and several new quads, courtyards, plazas, and arcades, with a restored stream offering shade and places for relaxation, socializing, and experiential learning for students and local residents alike.

Living systems

DISCIPLINE Campus Operations
ORGANIZATION The Society for College and University Planning

INDUSTRY Facilities
LOCATION Durham, New Hampshire, USA

Higher educational institutions in the United States regularly make land acknowledgments as part of their public events, recognizing that virtually every campus sits on land once occupied by Native American people. When New Hampshire's land grant university, the University of New Hampshire, engaged in a new planning effort, its planning team looked at the Indigenous history of the landscape and worked with representatives from the local Abenaki tribe. The team also engaged residents as well as faculty, staff, and students, using subject matter experts as well as Native American insights to develop a vision of the campus as a living system. They developed a map that highlights the ecosystems of the river and two brooks that weave through the campus and a public Campus Planner dashboard that provides information about everything from space use to demand projections to building performance. In addition, a decarbonization road map focuses on climate-related infrastructure. This is not just land acknowledgment but land action.

Mixed-use neighborhoods

DISCIPLINE Urban Planning   **INDUSTRY** Facilities

ORGANIZATION Jio Institute   **LOCATION** Maharashtra, India

The emergence of entirely new campuses, built from scratch, has become an increasingly common phenomenon, especially in rapidly developing nations. The Jio Institute is an excellent example of this: an entirely new campus that will occupy 400 acres in Navi Mumbai, India. Like Mumbai itself, the campus will have dense, mixed-use neighborhoods within walking distance of the academic, dining, and recreational amenities around a central green space. The campus will also have several facilities serving the larger community, including a research park, incubation center, convention center, a medical school and hospital, a K–12 school, and convenience stores. Developed during the COVID-19 pandemic, the design benefited from a global advisory group of experts from other universities to ensure that this plan met best practices in eight topic areas, from the arts to medicine. Mixed-use campuses such as this are like cities unto themselves, with the vertical layers of different activities in each building like the layers of spatial data in the geospatial tools used to develop them.

The Jio Institute in Navi Mumbai represents a global trend toward constructing new campuses from scratch to meet the substantial demand for higher education in developing countries. *Courtesy of Jio Institute, Sasaki.*

Regional systems

DISCIPLINE Urban Planning

ORGANIZATION The Society for College and University Planning

INDUSTRY Local Government

LOCATION Hanover, New Hampshire, USA

Universities often have a major impact on their regions not only as major employers but also as major landowners, which demands engagement of communities far beyond those of faculty, staff, and students. Dartmouth launched an 18-month comprehensive planning process for the 38,000 acres and six-mile-long corridor that it controls in and around its core campus. The 250-year-old university looked ahead to the next 250 years with a plan that conceives of the campus as a regional network of nodes, with a Catalog of Options that identifies future projects that could respond to the changing pedagogies, technologies, and fiscal realities facing higher education. The catalog also maps possible sites of new student, staff, and faculty housing units within walking distance of the main campus. In places where there are a lot of political boundaries and barriers to cooperation, colleges and universities can play a key role in getting those in power to think and act more regionally.

Dartmouth has completed a plan that connects its various campuses into a regional network of educational and research nodes. *Courtesy of Dartmouth College.*

Connections

DISCIPLINE Campus Operations   **INDUSTRY** Facilities

ORGANIZATION University of Minnesota   **LOCATION** Minneapolis, Minnesota, USA

Urban universities remain a part of the larger cities in which they stand, which affects access points and circulation patterns in and around the campus. The University of Minnesota, located at the center of the Twin Cities, along the Mississippi River, hired Sasaki to create a campus plan that looked at sites for new buildings and open spaces, mixed-mobility corridors, and gateways to the surrounding city. Each of the three distinct districts of the nearly 1,300-acre campus has its own character, with the West Bank campus requiring the renewal of its modernist podium plaza, the East Bank campus needing the reinforcement of mobility corridors and open spaces, and the St. Paul campus having to establish new open spaces and restore the existing landscape. The plan prioritizes the strengthening of campus gateways and the enhancement of transportation options, serving as a model of mobility in the larger city.

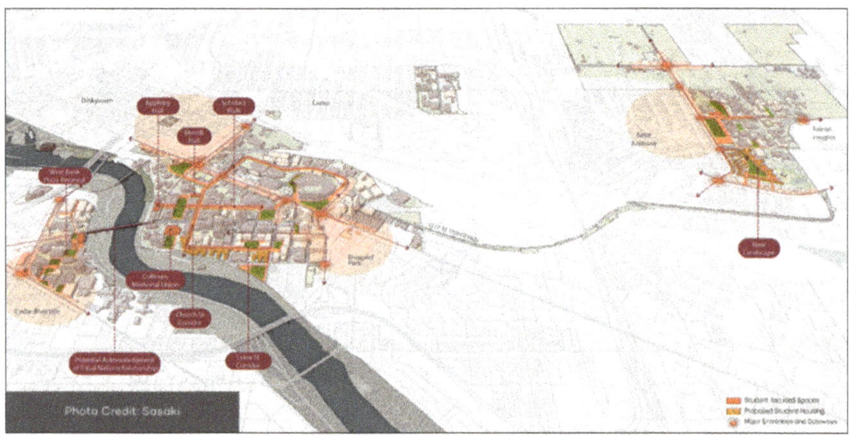

The University of Minnesota developed a campus plan that highlights mixed-mobility options, circulation corridors, and gateways to the rest of the Twin Cities. *Courtesy of the University of Minnesota, Sasaki.*

References

Accessibility: https://www.scup.org/award-winner/uc-berkeley-accessible-paths-and-places-master-plan/

Arizona State University: https://www.asu.edu

Asset Mapping: https://storymaps.arcgis.com/stories/e7e4301e045a4242bef57a66b5e1819d

Auraria Higher Education Center: https://aurariacampus.edu

Boston College: https://www.bc.edu

Connections: https://psre.umn.edu/sites/psre.umn.edu/files/2022-04/UMTC%20 Campus%20Plan%202021%20web.pdf

Dashboards: https://www.scup.org/award-winner/george-mason-university-george-mason-university-master-plan/

Digital twin: https://community.esri.com/t5/campus-operations-blog/integrated-campus-operations-the-digital-twin/ba-p/1503943

Emergency management: https://maps.usf.edu/ICM/

Expansions: https://www.scup.org/award-winner/the-university-of-the-west-indies-five-islands-campus/

Food: https://storymaps.arcgis.com/stories/9fac3f960a1f4fb7aecb1ccdefff32f8

Frameworks: https://www.scup.org/award-winner/emory-university-emory-university-framework-plan/

George Mason University: https://www.gmu.edu

Georgia Tech: https://www.gatech.edu

Health: https://www.uclahealth.org/sustainability/our-progress/sustainability-dashboard

Inclusivity: https://it.umich.edu/community/michigan-it-symposium/2020/posters /mapping-inclusivity-creating-crowdsourced-map-campus

Indoor Wayfinding: https://www.esri.com/en-us/lg/industry/education/german-universities-explore-indoor-mapping-system

Jio Institute: https://www.jioinstitute.edu.in

LandTech: https://www.landtechinc.com

Living systems: https://www.scup.org/award-winner/university-of-new-hampshire-university-of-new-hampshire-campus-as-a-living-system/

Mental health: https://gismaps.oit.ncsu.edu/portal/apps/experiencebuilder/experience /?id=755b9454351a43c6a408f0aec4f0b623&page=All-Greenspaces

Michigan State University: https://msu.edu

Mississippi Watershed Management Organization: https://www.mwmo.org

Mixed-use neighborhoods: https://www.jioinstitute.edu.in/sites/default/files/media_ document/Jio%20Institute%20Introduction%20Presentation.pdf

Navigation: https://community.esri.com/t5/campus-operations-blog/how-the-university-of-texas-at-austin-is/ba-p/1497849

NC State University: https://www.ncsu.edu

Ohio State University: https://www.osu.edu

Parking: https://facilities.gatech.edu/sites/default/files/2023-05/Parking_and_ Transportation_Demand_Management_Immediacy_0.pdf

Public art: https://storymaps.arcgis.com/stories/b9a60162175f4109b17b770561385a97

Regional systems: https://www.scup.org/award-winner/dartmouth-planning-for-possibilities-a-strategic-campus-framework/

Shared facilities: https://www.scup.org/award-winner/aurarua-higher-education-center-auraria-campus-framework-plan/

Space efficiency: https://www2.cortland.edu/offices/facilities-management/20180529%20 SUNY%20Cortland%20FMP%20Update.pdf

Space management: https://community.esri.com/t5/campus-operations-blog/how-asu-leverages-gis-to-map-and-manage-its/ba-p/1574878

St. Mary's University: https://www.smumn.edu

SUNY Cortland: https://www2.cortland.edu

The Society for College and University Planning: https://www.scup.org
Theft: https://www.landtechinc.com/wp-content/uploads/2022/06/LandTech_CaseStudy_
    UMass_Lowell_ArcGIS_Indoors_final.pdf
Trash: https://storymaps.arcgis.com/stories/5d2133a555e7456c9f95b7181bee3e42/print
Tree canopy: https://storymaps.arcgis.com/stories/57ab71662f64488db86b5ba3c4f30dfa
UC Santa Barbara: https://www.ucsb.edu
UCLA Health: https://www.uclahealth.org
University of Applied Sciences in Dresden: https://www.htw-dresden.de
University of California, Berkeley: https://www.berkeley.edu
University of Kentucky: https://www.uky.edu
University of Michigan: https://umich.edu
University of Minnesota: https://umn.edu
University of Redlands: https://www.redlands.edu
University of South Florida: https://usf.edu
University of Texas at Austin: https://www.utexas.edu
University of the West Indies: https://www.uwi.edu
Utilities: https://www.arcgis.com/apps/webappviewer/index.html?id=
    f378ea1bbbe34a2eae12793d24f08c66
Visions: https://www.scup.org/award-winner/sacramento-state-placer-center-sacramento-
    state-placer-center-master-plan/
Water: https://www.mwmo.org/path-to-the-river-main/
Wayfinding: https://experience.arcgis.com/experience
    /b541128d07a64be78257ffda93b2dc73/page/Accessibility-Map/

Credits and contributions

Sue Stewart, assistant director for Campus Geospatial Assets, the University of Texas
    at Austin
Sean Moran, GIS professor, Austin Community College
Brian Cox, GIS intern, Austin Community College
Kelsey Grant, GIS intern, Austin Community College
Dartmouth College board of trustees, executive leadership, campus planning, and dozens
    of other administrators, as well as the thousands of students, staff, faculty, and regional
    community and business members who participated in the planning
  - Beyer Blinder Belle Architects & Planners
  - Michael Van Valkenburgh Associates
  - BFJ Planning; Nitsch Engineering
  - Atelier Ten
  - BuroHappold
Sasaki; RMF Engineering; Kimley-Horn; Walker Consultants; Building Conservation
    Associates; Pattern R&D
Sasaki; Arup; Sherwood Design Engineers; Mark Cavagnero Associates; U3 Advisors;
    StudioUmmo

CHAPTER 5

Community engagement

Introduction

Most colleges and universities have a commitment to engaging communities, be they the communities where the higher education institutions are located or the broader regional, national, or even global communities of which they—and almost everyone else—are a part. Geospatial tools—GIS—offer a way to engage and empower these communities: to help them see what's happening around them and enable them to envision a better future for themselves. Maps help communities make sense of the world and, at the same time, give them hope.

Institutions of higher education approach community engagement in different ways. Some coordinate it centrally, with a community engagement office that connects faculty, staff, and students to opportunities, often in the neighborhoods close to campus or at a location with some connection to the institution. Others take a more decentralized approach, encouraging individual faculty members or research center directors to take the lead with community engagement, whether locally, nationally, or internationally, depending on the funding of the work or the goals of a course.

In all cases, GIS offers a useful tool in understanding communities. Its ability to juxtapose diverse datasets to see unseen relationships and draw unexpected conclusions makes it an excellent way of helping community members and their partners see their places in new ways. GIS also allows us to do asset mapping, allowing community members to identify the elements of a place that matter the most to them, letting them secure the future they want. Maps become a way of helping neighbors, despite their differences, recognize what they have in common and what goals they share.

The following examples describe what that engagement looks like at a variety of scales. These maps and case studies indicate the depth and breadth of higher education's involvement in communities of all kinds, near and far. We have sorted these projects according to their scale, starting with examples addressing global communities, followed by those at a national, regional, and local scale. There are more of the latter because a lot of the available data underpinning these maps exists at the local and regional scale, in the databases of city, county, or state or provincial governments. Nevertheless, through maps, communities at all scales make major contributions to our understanding of the world. This introduction provides an overview of the case studies that follow, each of which discusses in greater detail how GIS is being used in communities at a range of scales.

Global communities

Global communities may be the hardest to define, considering their enormous scope and diversity, but they may also be the most important ones, considering the global ecological challenges that humanity now faces. Researchers at Esri and the United States Geological Survey (USGS), along with a World Terrestrial Ecosystems app developed at the University of Minnesota, have mapped the globe's ecosystems, defining the planet not by national boundaries but by ecosystems we share. We also share the oceanic areas, which make up more than 70 percent of the planet's surface and contain more than 50 percent of life on earth. And unlike the most threatened terrestrial ecosystems, the marine-based ecosystems in need of protection are unevenly distributed, with coastal areas and large patches in the middle of oceans as the most imperiled.

Maps can sometimes document themes or trends that apply across the globe. Mapping the news, for instance, may address issues that apply in many places, having to do with conflicts among nation-states, impacts of public policies, and changes to the global climate or challenges to public health. Some global maps may also not have a geographic character at all. Knowledge maps, such as a map showing clusters of professionals who engage in participatory planning and the influence of some planners on others, indicate the strength and scope of interactions among people or things, regardless of their physical location. These intellectual geographies help us not only visualize the often-invisible nature of human connections but also imagine relationships free of the boundaries that often impede professional or personal relationships.

National communities

Some communities include professional collaborations, such as the International Geodesign Collaboration, a network of more than 240 universities that has produced numerous maps, including a climate action map of Portugal. Maps can build communities among those who make maps. They can also reveal the connections among people at a national level, as described by the case study of citizen scientists and their work in European countries or the one that locates GIS programs in the United States. Sometimes the best work within nations arises from a comparison with others, which maps allow.

Climate change, however, may present the greatest challenge for nations, driven by the accumulation of atmospheric carbon. The study of how England might deal with that challenge shows that it demands a variety of strategies involving several sectors, from agriculture and forestry to transportation and energy production. And in parts of the world, water scarcity may present an even greater challenge. The case study of water scarcity in South Africa describes how nations need to address inequities within their borders, including where they can sustain their populations and where they can't. All of this demands political will, something that election maps of the United States reveal when a country becomes so divided that making progress on addressing climate change can present a problem.

Regional communities

In such cases, the greatest progress may happen at the regional scale, within subnational governments or local landscapes. Election redistricting can reduce political division or, as the redistricting study of Washington state suggests, make those divisions more pronounced. Maps can also reveal the impact of inaction, as the study of how development in Connecticut has swallowed up prime agricultural land reveals, leading to needed policy discussions. Also, the more closely a map focuses on our own regions, the harder it is to avoid responsibility, as the study of responsible land management of the Winooski River watershed in Vermont describes.

Maps not only show us where to go; they also show us why we might want to go in one direction or another. The description of COVID-19 risks in Vancouver shows where we need to be better prepared in case of a future pandemic, and the description of opioid abuse in Tennessee indicates where we need to focus our attention in curbing drug addiction. Maps can also help us predict future problems, as the case of greater-than-expected water demand around Chicago indicates. Effective maps elicit not just an "oh" but an "uh-oh," and the best maps remind us of what we have lost, eliciting an "oh no." The study of the Indigenous landscape of the Los Angeles basin shows how native populations saw landscapes as a whole, rather than divided into private properties, and the map of the Indigenous landscape in northern Minnesota shows how human settlements and travel paths once had—and could have again—a minimal impact on the land.

Minimizing human impacts is especially pertinent on islands, which often have unique ecologies because of their isolation from the mainland and greater exposure than most of the mainland to rising sea levels. The study of how the island of Sardinia might respond to climate change demonstrates how island communities can sometimes collaborate in ways that some mainland decision-makers do not, looking at how every part of the island plays a part in the solution. Meanwhile, Harvard's study of the island of Puerto Rico validates the role that higher education institutions can play in generating innovative responses to climate change.

Local communities

Universities, however, present problems as well as solutions, evident in the case of the University of Georgia's displacement of a Black community to build dormitories many decades ago. At the same time, universities can reveal the inequities within the communities around them, as the maps of the impact of racial and ethnic redlining in California's Inland Empire reveal. Owning up to past prejudices is key to mapping a more equitable future.

One way to achieve greater equity is by engaging citizens in the making of maps. The studies of how the greening of vacant sites in Philadelphia has improved adjacent land value or how diverse populations do not feel welcome in the open spaces of Copenhagen's Amager district involved a lot of public participation. And citizen science drove the documentation of feral plant life in Dublin and air quality in London, showing how GIS technology has enabled people to help make maps as never before.

The threats of climate change can engage the public in new ways, whether it's the study of the damage of coastal flooding in Winneba, Ghana, or the risk of a tsunami in coastal Washington state. The goal of economic prosperity can also attract public participation. The study of how to balance the economy and ecology of Huzhou, China, offers one example of that, and the study of how to balance vehicular and pedestrian access in Tempe, Arizona, offers another. In the following examples, mapping becomes a way to build, reveal, and improve communities.

Global community examples

Mapping the news

DISCIPLINE Journalism		**INDUSTRY** Business	
ORGANIZATION University of Wisconsin—Milwaukee		**LOCATION** Global	

Print media have long used maps to convey the news in a visually and geographically compelling way, and universities often have some of the largest collections of such maps. The University of Wisconsin-Milwaukee's American Geographical Society Library is an example. Its 2024 installment of the *Maps and America* exhibit included a lecture by Tim Wallace, the senior editor for geography at the *New York Times*, titled "Newsroom Cartography," and a story map that looked at how print media has used maps over the last century or more to convey the news related to world affairs, wars, elections, local interests, and climate change. "Newsroom Cartography" has told stories in colorful and compelling ways, sometimes sacrificing the accuracy of the geography to say what needs to be said about a topic. In print media, it's the point rather than the preciseness of the map that matters most.

Mapping world ecosystems

DISCIPLINE Ecology

ORGANIZATION University of Minnesota
GeoCommunities

INDUSTRY Sustainable Development

LOCATION Global

We have become so accustomed to seeing the world in terms of national and continental boundaries that we often forget the ecosystems that support us and all the other living organisms on the planet. The Connected Through Climate app seeks to change how we think about the earth. Developed by GeoCommunities at the University of Minnesota and using the *World Terrestrial Ecosystems* map created by Esri and the USGS, the app lists the world's 18 climate-based ecosystems, arranged by size, and it maps those ecosystems on an interactive globe. The app allows users to turn on or off individual ecosystems, as well as cities and national borders, and it shows how ecosystems will change between now and 2050 in terms of heat, aridity, and land cover. Dashboards for each of the ecosystems allow communities occupying the same ecosystem to share information and best practices, regardless of their political differences. And dashboards allow an analysis of public policies in various countries, which can have dramatically different effects on ecosystems. You could call it an apt app.

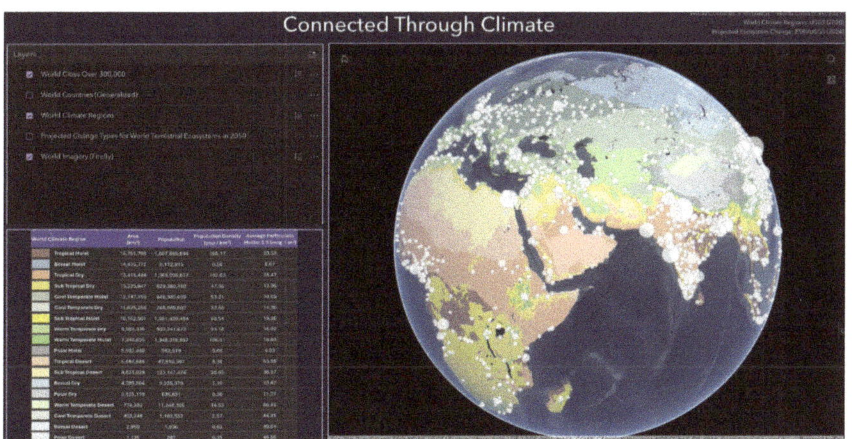

This map of world terrestrial ecosystems by the University of Minnesota shows how ecosystems vary widely while also connecting disparate parts of the world. *Courtesy of Esri, USGS, U-Spatial at the University of Minnesota.*

Protecting ocean biodiversity

DISCIPLINE Oceanography

ORGANIZATION National Geographic Society

INDUSTRY Science

LOCATION Global

The oceans occupy more than 70 percent of the earth's surface and contain more than 50 percent of its living matter, and yet these bodies of water often get less attention than the land when it comes to the protection of biodiversity. This map of the world's oceans highlights in gradations of color, from the lowest-priority areas in blue to higher-priority areas in increasing shades of yellow, where the greatest oceanic biodiversity exists and where we need to focus our protection efforts. Based on the research of faculty from a number of universities, including Oregon State and the University of California, Santa Cruz, and coordinated by the National Geographic Society, the map locates the Marine Protected Areas (MPAs), most of which exist along developed coastlines and in large patches in the Pacific, Indian, and Arctic Oceans. The MPAs show where altering "abatable impacts" by human life would lead to marginal gains in the persistence of biodiversity in the world's oceans. This fixes our sights on what we need to do in that highly fluid world.

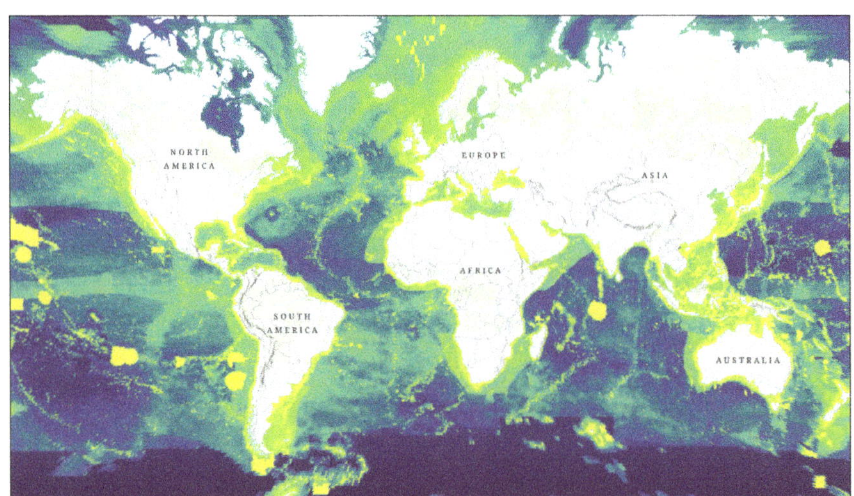

This map shows where oceanic biodiversity is most threatened and where protective measures are most needed. *Courtesy of ArcGIS Living Atlas of the World.*

Identifying planners

DISCIPLINE Data Science   **INDUSTRY** Education

ORGANIZATION University of Lodz   **LOCATION** Global

The use of GIS in public participation planning has grown since the 1990s, although this community of practitioners and researchers has not mapped itself—until now. A faculty member at both the University of Lodz, Poland, and Al-Zahar University in Egypt analyzed the metadata of 699 published articles about public participation geographic information systems (PPGIS) from 1997 to 2023.[1] The literature fell into 20 thematic categories, and some topics, such as land planning and management, environmental quality, and climate change vulnerability, remained underrepresented. The analysis also found that researchers in the United States, Finland, Australia, and the United Kingdom dominated the output of articles, with an annual growth rate in the number of articles of 8.9 percent. Mapping the coauthors of articles also showed clusters of researchers, with a few key contributors and their collaborators highly linked; the size of each circle indicated their publication count, and the thickness of the lines indicated the strength of their connections. Such mind maps have an astronomical character, reminding us that universities and the universe have the same root word: *universum*, meaning combined into one whole.

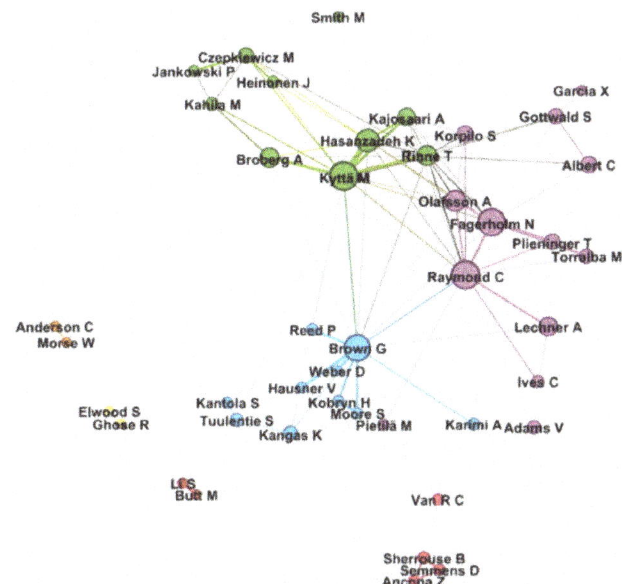

Knowledge maps have emerged as an important way of understanding the relationships among people, ideas, or units in a given field or organization, as this knowledge map of GIS researchers shows. *Courtesy of Abdelbaseer A. Mohamed.*

National community examples

Going from global to national

DISCIPLINE GIS

INDUSTRY Sustainable Development

ORGANIZATION University of Lisbon

LOCATION Portugal

When it comes to actions aimed at curbing humans' impact on the global climate, the timing and sequencing of those actions can matter as much as the actions themselves. This pilot project, led by researchers at the University of Lisbon and colleagues in the International Geodesign Collaboration (IGC), looked not only at where a set of climate actions across several domains—energy, transportation, agriculture, forestry, and so on—could occur in Portugal but also at when and in what order those actions would have the greatest impact. The research team, working with students and community stakeholders, developed a mapping strategy that allows people to visualize overlapping climate actions at the same location and a timing strategy that lets users assign start and end dates for each action and then arrange them chronologically to aid decision-makers in determining which policies and processes need to take precedence. This approach, launched as part of the IGC's Global Climate Geodesign Challenge, holds great promise in dealing with climate mitigation, helping us know not only what to do but also where and when to do it.

Providing election information in the United States

DISCIPLINE Political Science

INDUSTRY Government

ORGANIZATION University of Florida Election Lab

LOCATION Florida, USA

Colleges and universities play a key role in the life of their countries in gathering and sharing data about elections, serving as a source of information that isn't compromised by political spin. The University of Florida's Election Lab serves that role as a trusted source of statistical and geospatial information about elections. The lab has precinct-level election results in all 50 states, going back several years, with maps accompanying many of them. The database shows how closely divided most statewide elections are in terms of the number of voters and how skewed they are spatially, with the Republican Party dominating rural areas and the Democratic Party metro areas. But "blue" America shouldn't necessarily be blue about these mostly red maps; the urban-rural divide seen in American politics has become the case in many countries around the world.

Locating citizen scientists in Europe

DISCIPLINE GIS

ORGANIZATION European Citizen Science Association (ECSA)

INDUSTRY Education

LOCATION Europe

Although universities and their faculty, staff, and students dominate the research community globally, citizen science has a long history, with some of the greatest discoveries in science coming from people not connected to a university. The European Citizen Science (ECS) program, with university partners that include University College London, the University of Leiden, and the University of Coimbra, supports citizen science efforts in countries around the world, with a concentration in Europe. This interactive map allows interested parties to see who is working on what and how to access their work, with connections to websites and descriptions of projects. The map also shows the institutions that support citizen science. The great challenge of citizen scientists is their isolation, in contrast to university research with its framework of departments, centers, and peer-reviewed publications providing places for collaborators to gather and share work. ECS offers a similar gathering place for citizen scientists without those other supports, with maps to help people find their way—as maps often do.

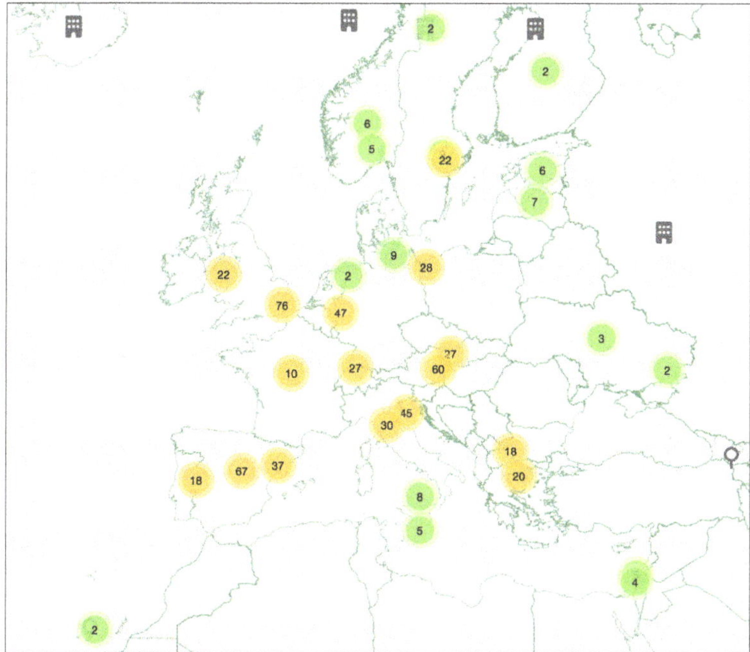

This map of participants across Europe shows the scope of the European Citizen Science platform in building the capacity and awareness of citizen scientists. *Courtesy of European Citizen Science. Copyright © citizenscience.eu, CC BY 4.0.*

Planning for climate change in England

DISCIPLINE Environmental Science   **INDUSTRY** Sustainable Development

ORGANIZATION University of Manchester   **LOCATION** United Kingdom

Many countries have wisely decided to plan for the impact of climate change on their land and seas, and that includes England, which has set ambitious carbon-reduction goals and focused on clean growth, green industrial strategies, and climate adaptations. This study by a research team at the University of Manchester looks at climate actions in the eight domains of the Global Geodesign Challenge: energy, agriculture, forests, oceans, settlements, industry, transport, and water. And they examined a wide range of solutions, including locating solar farms on less fertile land, protecting peatlands, restoring grasslands and salt marshes, renovating buildings to be zero-carbon, decarbonizing industrial facilities, constructing high-speed rail, and restoring river ecosystems. The geodesign team commends the government for its goal of being net zero by 2050 but notes that the plans have no spatial component, which this study does a good job of remedying. What a government wants to do has little value unless it can also identify what it needs to do where. Policy without geography can only go so far.

Addressing climate change across an entire nation requires action on all fronts and in every sector to succeed. *Courtesy of Richard Kingston, University of Manchester.*

Highlighting water scarcity in South Africa

DISCIPLINE Environmental Science **INDUSTRY** Government

ORGANIZATION University of Edinburgh **LOCATION** South Africa

There may be no greater threat to a population than lack of access to water, making the mapping of water scarcity a critically important issue. This GIS study of South Africa by a diverse group of researchers at the University of Edinburgh shows how a majority of households lack access to an adequate water supply in the western half of the nation and many areas in its eastern half. Such maps helped prompt the creation of a National Spatial Development Framework as the basis for long-range planning and for bringing together many agencies, each holding its own data with little coordination among them. South Africa also created a central map library of spatial information that supported the national framework for data management. Since then, the country has begun to address the needs of its most underserved population in a country that still has one of the largest income gaps in the world between White and Black inhabitants. GIS not only has the power to reveal the most important challenges facing a population but also to integrate data and connect decision-makers in ways that work for everyone. Data is like water: it needs to be accessible, and it endangers us when it's scarce.

Finding GIS programs in the United States

DISCIPLINE GIS **INDUSTRY** Education

ORGANIZATION GeoTech Center **LOCATION** USA

Prospective students interested in mapping would look to a map to identify the academic programs that offer relevant degrees, and this case provides such a location tool. The map was developed by researchers at the National GeoTech Center of Excellence, a National Science Foundation Center at the Jefferson Community and Technical College in Louisville, Kentucky. The organization provides more than maps. It also operates a Geospatial Education Awards program, research studies that show where GIS-related jobs are most common, continuing education conferences and certificate programs, and a self-assessment tool to help prospective employees identify workplace-relevant skills. Maps can make a difference for everyone but especially those who want to make maps.

Regional community examples

Redistricting statewide elections

DISCIPLINE Political Science   **INDUSTRY** Local Government

ORGANIZATION League of Women Voters of   **LOCATION** Washington, USA
Washington

The use of redistricting to gerrymander political boundaries for political advantage of one party or another has become a fierce battle in the United States. In response to that situation, the League of Women Voters of Washington and the group Politics of the Possible in Action worked with Integral GIS to create a series of redistricting maps that better represent the state's population and the will of its voters. Called the Washington Redistricting Lenses project, the project with the University of Washington used demographic and spatial data from the most recent American Community Survey to generate maps that accounted for the interests of many groups, according to race/ethnicity, household income, tribal affiliations, environment, and industry employment. The researchers then used ArcGIS to describe the maps and their generation and highlight how redistricting can be used to favor one group over another or to create more equitable and competitive districts that make everyone's vote count.

Recognizing Indigenous communities

DISCIPLINE History   **INDUSTRY** Education

ORGANIZATION UCLA   **LOCATION** Los Angeles, California, USA

Los Angeles has the largest Indigenous urban population in the United States, and yet little mapping of that community has been done until now, when a team of UCLA researchers and students worked with community members to develop a series of story maps about the Indigenous experience in the LA basin. Rather than tell a chronological story about the arrival of colonizers and the disappearance of the native Tongva people, the story maps tell a more complex story in which the area's Indigenous people continue to occupy the land and thrive as a community in the region. Historical maps, drawings, and photographs reveal how the land was taken from its native inhabitants and how they fought to maintain their identity and teach others about it, with annual events such as the Life Before Columbus Festival. We all have much to learn from native communities about how to care for the land, and story maps are a good start.

Reducing climate impacts

DISCIPLINE Environmental Science   **INDUSTRY** Sustainable Development

ORGANIZATION University of Vermont   **LOCATION** Vermont, United States

Climate change may have some of its greatest impacts not only along vulnerable coasts but also inland, where forests and fields can suffer from higher temperatures and greater aridity. This study of the Winooski River watershed in Vermont by a research team at the University of Vermont looked at several issues, including renewable energy opportunities, wastewater efficiency improvements, river ecosystem restoration, vehicle electrification, and natural solutions to water quality, all of which can mitigate climate impacts. The team also looked at land trusts as a way to preserve the watershed's rural character, and they focused on the transportation and energy sectors, which contribute the most to climate change. The final design shows how all these solutions need to work in concert to achieve the climate goals we all need to pursue. Maps allow us to see the whole, even if that whole involves a lot of different and important parts.

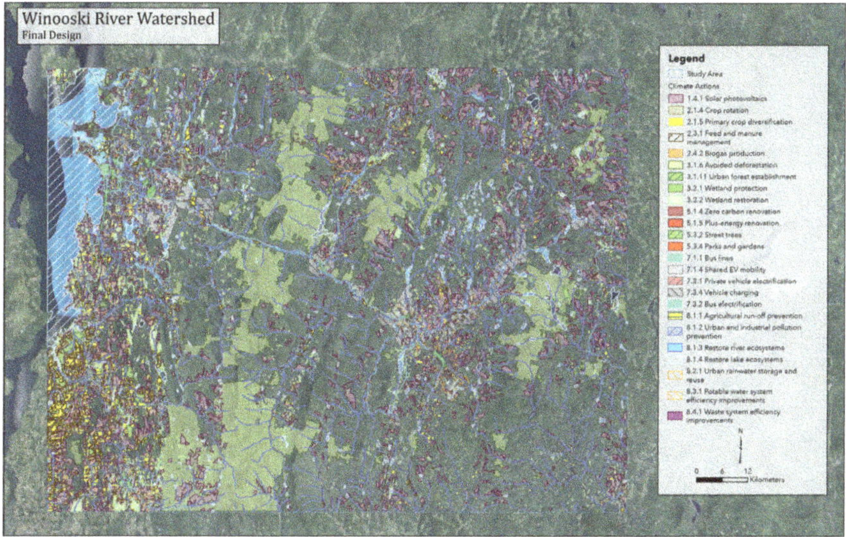

Reducing the impact of climate change through renewable energy projects requires different strategies, as this map of Vermont's Winooski River watershed suggests. *Courtesy of the University of Vermont.*

Anticipating COVID-19 risks

DISCIPLINE Public Health

ORGANIZATION Simon Fraser University

INDUSTRY Health and Human Services

LOCATION Vancouver, British Columbia, Canada

The dashboard developed by Johns Hopkins University to track the rate and spread of the COVID-19 virus during the pandemic became the go-to source of information for many people around the world. But this story map by a research team at Simon Fraser University took a finer-grained approach and analyzed differences in personal behavior and neighborhood environments in the region in and around Vancouver, British Columbia, Canada, to produce an overall virus risk map. It showed how a virus might spread in unequal ways, with some neighborhoods having a greater density of shared indoor spaces, overcrowded housing, transit dependence, and in-person work, among other factors, which can increase the risk that people would contract a virus such as COVID-19. As we think about how to prepare for future pandemics, we should focus on those places where the risks of becoming ill are highest and where the rewards for reducing risk factors are greatest.

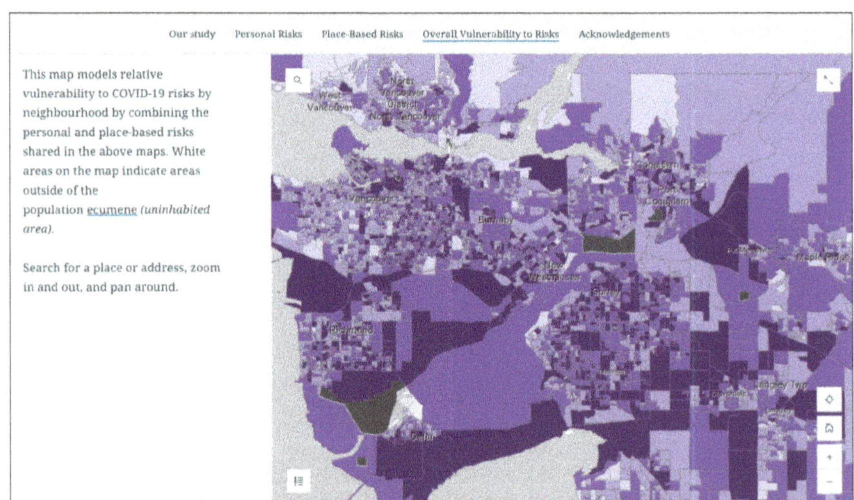

A pandemic such as COVID-19 affects everyone, but not equally. This map of the risk of catching COVID-19 showed variations by neighborhood, based on an analysis of personal and place-based risks. *Courtesy of Simon Fraser University, Dr. Valorie Crooks, and Nadine Schuurman.*

Recovering the Indigenous landscape

DISCIPLINE History

INDUSTRY Sustainable Development

ORGANIZATION Minnesota Design Center

LOCATION Minnesota, USA

Native landscapes evolved to thrive in particular ecosystems and absorb atmospheric carbon at a remarkable rate. Created as part of the International Geodesign Collaboration's Global Climate Geodesign Challenge, this map by the University of Minnesota arose from an examination of the native landscape in Northern Minnesota and its carbon-storing capacity, with the goal of seeing how far we could go in returning as much of the current landscape as possible to that native state. Using the GIS-based app developed as part of the challenge, the research team imagined North America as if the 17th-century French idea of trading with the native community had prevailed and the Indigenous environment had survived. Based on that idea, the team assessed the carbon sequestration impact of having compact trading posts, connected by rail lines, in an otherwise native landscape. Calculating the carbon sequestration capacity of the Indigenous landscape suggests that the more we can return the land to native plant communities, the better we—and the atmosphere—will be.

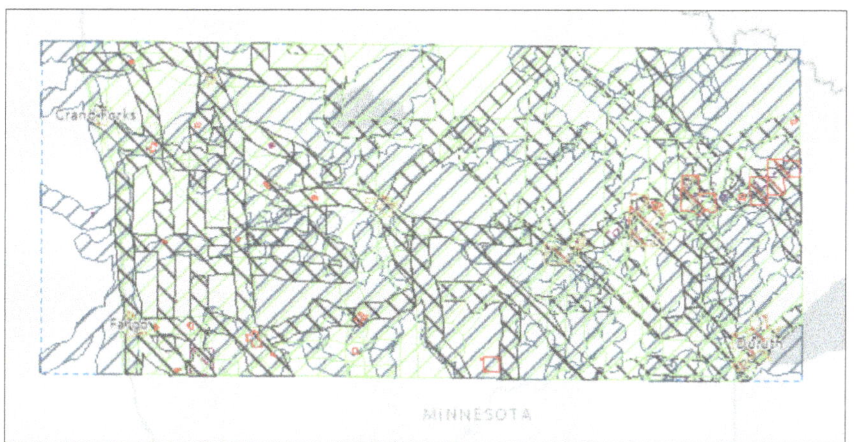

If we hope to sequester atmospheric carbon to reduce climate change, we need to rediscover and restore the carbon-absorbing capacity of Indigenous landscapes, as in this Northern Minnesota example. *Courtesy of Minnesota Design Center, University of Minnesota.*

Decarbonizing Sardinia

DISCIPLINE Environmental Science   **INDUSTRY** Sustainable Development

ORGANIZATION Cagliari University   **LOCATION** Sardinia, Italy

Islands have the advantage of clear borders and a sense that everyone needs to work together to thrive. The work of a research team from Cagliari University shows what an island such as Sardinia can achieve when people share information and work in concert. The team led workshops that included faculty and students as well as representatives from 17 municipalities and the metropolitan authority; they looked at the eight Geodesign Challenge domains, plus tourism and cultural heritage. They based their work on objectives such as sustainable transportation, green energy and infrastructure, protection of natural areas and wetlands, protection from flooding, improved agriculture and water quality, and the creation of cultural heritage networks and sustainable tourism. And they produced short-, medium-, and long-term strategies to decarbonize Sardinia that offer a road map that everyone can follow. In the end, no one is an island.

The island of Sardinia has identified several actions that will decarbonize the communities across their entire landscape. *Courtesy of Michele Campagna, Cagliari University.*

Predicting regional water demand

DISCIPLINE Environmental Science

ORGANIZATION University of Illinois Extension

INDUSTRY Sustainable Development

LOCATION Chicago, Illinois, USA

Access to water has become a vitally important challenge, not just in arid regions but also in those regions with access to ample freshwater, such as Northeastern Illinois on Lake Michigan. The University of Illinois Extension partnered with the Chicago Metropolitan Agency for Planning and Illinois-Indiana Sea Grant to update the forecast of the region's water demand. Although the study showed that, overall, water use has decreased because of advances in water conservation and efficiency, suburban and exurban areas in the greater Chicago region should experience an increase in water demand that is greater (in some places, by more than twofold) than the available groundwater supplies. This mapping of demand versus supply suggests that measures need to be taken as soon as possible to reduce water consumption to avoid future water scarcity, a challenge made more difficult with the building of data centers close to Chicago. These centers, which allow us to produce such data-based maps, also demand a lot of water.

Tracking opioid use

DISCIPLINE Public Health

ORGANIZATION University of Tennessee, Knoxville

INDUSTRY Health and Human Services

LOCATION Tennessee, USA

The opioid crisis has affected parts of the United States and counties within each state differently, as this data mapping project from the University of Tennessee describes. The author of the study used *The Washington Post* opioid dataset to look at the spatial distribution of buyers and sellers of opioids in Tennessee. The annual amount of opioid use in the state was astounding—about 360 pills per person. But mapping the data showed that, while the largest cities—Memphis, Nashville, and Knoxville—had the largest number of pills distributed, the number of pills taken per capita showed that the greatest amount of opioid abuse by individuals occurred in rural counties, far from cities. The author conjectures that this may be because those counties had fewer pharmacies that attracted people from surrounding counties, had lower prices than in the cities, or had more nursing homes and assisted living facilities per capita. Here, geospatial tools illustrate a topic that cuts across the urban-rural divide, representing a public-health disaster for the entire country.

Transforming Puerto Rico

DISCIPLINE Geodesign

INDUSTRY Urban Design

ORGANIZATION Harvard Graduate School of Design

LOCATION Puerto Rico

Universities have a lot to offer communities in terms of exploring new ideas in unthreatening ways. This study of Puerto Rico by faculty, staff, and students at Harvard University shows how valuable such academic projects can be. Aiming to reduce carbon emissions and increase carbon sequestration, the design team set goals for renewable energy, reforestation, carbon farming, coral reef restoration, wetland conservation, and public transportation. Based on those goals, students produced a number of innovative ideas, such as passive cooling spaces, biorefineries with sugarcane, air purification facades, urban forestry, air filtration billboards, phytotechnology, and offshore photobioreactors. The team plotted those strategies on a timeline and mapped them all to assess their collective contribution to reducing atmospheric carbon. This project gave Puerto Rico, which means "rich port," the ability to harbor a rich cache of new ideas.

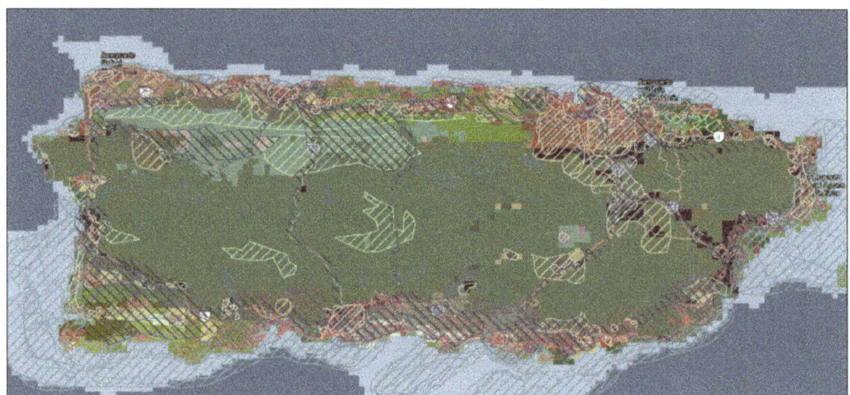

Reducing the carbon footprint of Puerto Rico requires a range of ambitious climate actions across the entire island. *Courtesy of Stephen Erwin, Carl Steinitz, Harvard University.*

Showing landscape change

DISCIPLINE Remote Sensing **INDUSTRY** Sustainable Development

ORGANIZATION University of Connecticut **LOCATION** Connecticut, USA

Sometimes we do pave paradise, in Joni Mitchell's words, and this study of *Connecticut's Changing Landscape* by the University of Connecticut makes that clear with a series of maps that show where development and turf grass covered the land between 1985 and 2010, where good agricultural soils have been lost to other uses, where vegetation along streams has been lost, and where tree canopies have declined. Accompanied by data and text that make a convincing case for stricter controls on development, the story map also gives a powerful example of how well-illustrated research data not only communicates the facts of a situation but also helps us understand how we might feel about the situation and what we might do about it. There are also pragmatic uses for studies like this, in helping a state or region understand what it's losing before it's too late.

Local community examples

Planning for tsunami evacuation

DISCIPLINE Public Health **INDUSTRY** Emergency Management

ORGANIZATION Washington Military Department/WA Emergency Management **LOCATION** Camp Murray, Washington, USA

Coastal communities around the Pacific Ring of Fire have to deal not only with seismic events but also the tsunamis triggered by those events, in some cases thousands of miles from the initial event. For example, a subduction zone earthquake in the Aleutian Islands may generate a tsunami with enough energy to cross the Pacific and impact shores in Chile or Japan in less than one day. A team from the University of Washington worked with the Office of Emergency Management in Grays Harbor County, Washington, to identify suitable areas for residents to evacuate to if a tsunami hit is expected. The GIS analysis showed that there isn't enough public land to accommodate the number of evacuees. So the research team looked at private land that could serve as evacuation sites. They also looked at areas that are suitable, given enough evacuation warning, versus spaces closer to population centers that may need to handle large numbers of evacuees very quickly. Geospatial tools enable such analyses in ways that give communities options when faced with little space or time to react to threats such as tsunamis.

Uncovering university-driven displacement

DISCIPLINE Urban Planning   **INDUSTRY** Urban Design
ORGANIZATION University of Georgia   **LOCATION** Athens, Georgia, USA

We often think of urban renewal in the 1960s occurring in the center of large cities, but large universities also engaged in their share of displacement. The Linnentown neighborhood in Athens, Georgia, was a Black, working-class community just west of the University of Georgia's campus. In 1961, the university had the City of Athens designate the community as an urban renewal site, without notifying the residents living there, and then purchased and demolished their houses to make room for large, high-rise dormitories, parking lots, and a parking deck. The Community Mapping Lab at the university has documented the history of this neighborhood's erasure in a publicly accessible story map that documents the history of the community, contains a virtual walking tour of the site, and recounts the residents' resistance to the community's elimination. Students and faculty also launched a public art and education project to help the University of Georgia community and the city of Athens reflect on what happened to Linnentown. Maps help us remember what some might want to forget.

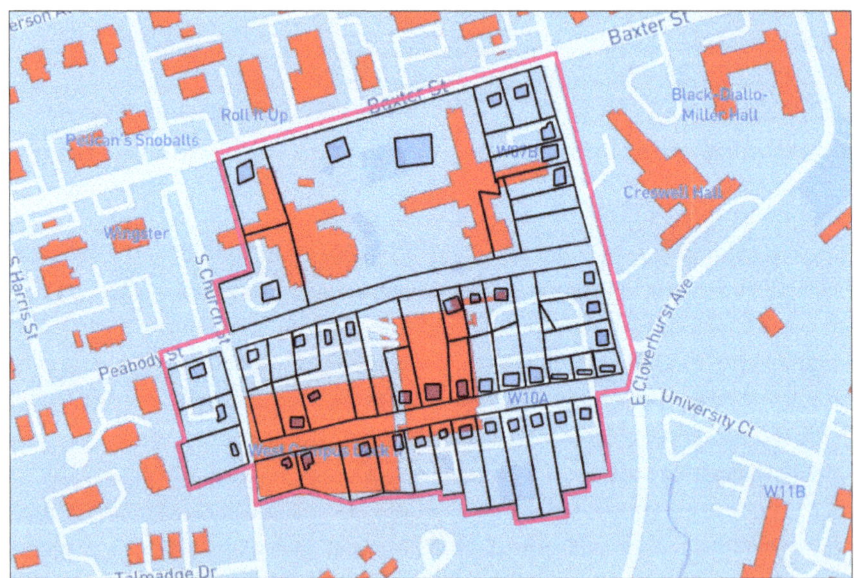

This map of the Linnentown neighborhood in Athens, Georgia, shows the parcels purchased and the buildings demolished by the University of Georgia to make room for new dormitories.

Courtesy of the Community Mapping Lab, University of Georgia, in consultation with Linnentown residents.

Revealing redlining

DISCIPLINE History

ORGANIZATION University of Redlands

INDUSTRY Urban Design

LOCATION California, USA

Federal law in the United States prohibited discrimination in residential lending, known as redlining, in 1968, but the results of that now-illegal practice remain evident in cities to this day. The People's History of the Inland Empire project involves local universities, historical societies, libraries, and archives in documenting the histories of communities of color, working people, and LGBTQ+ people in California's Riverside and San Bernadino Counties through maps and story maps. This redlining map exemplifies the group's work. The map documents where diverse racial or ethnic categories of households lived in each decade of the 20th and early 21st century, according to the US Census. Using Esri's *Human Geography Dark Map* as the basemap, researchers at the Program in Race and Ethnic Studies at the University of Redlands Center for Spatial Studies traced the effect of redlining on the Hispanic community in the Inland Empire. Confined to a few tightly defined neighborhoods, the Hispanic population shows up as bright-red blocks amid the green dots of the non-Hispanic households. The ongoing concentration of poverty in those red zones serves as a reminder that the work to counter the effects of redlining is not yet done.

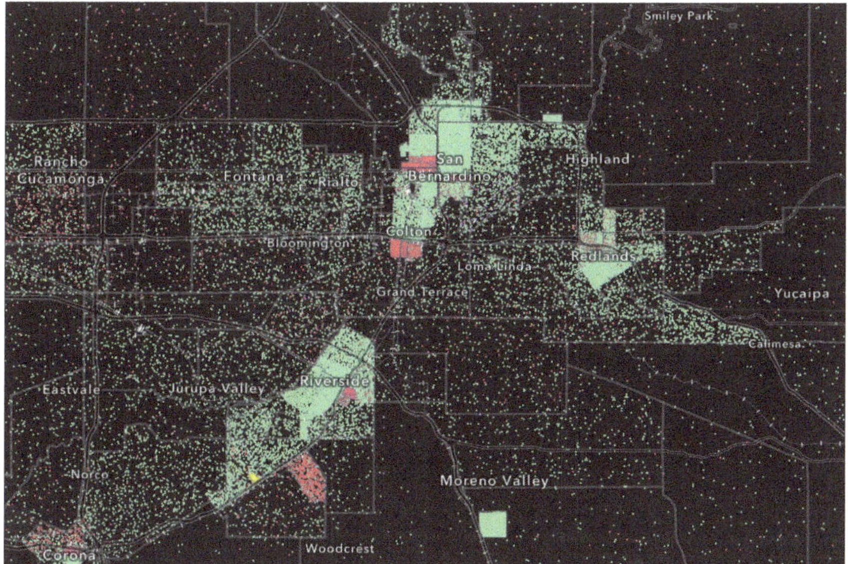

Redlining became illegal in 1968, but its legacy continues in communities that have concentrated areas of poverty, as this map of California's Inland Empire shows. *Courtesy of the People's History of the Inland Empire Census Project.*

Planning for recognition justice

DISCIPLINE Public Policy   **INDUSTRY** Urban Design

ORGANIZATION University of Helsinki   **LOCATION** Copenhagen, Denmark

Public participation in planning has become an important part of many planning efforts, although marginalized groups often remain underrepresented. To address that challenge, a group of researchers from the University of Helsinki and the University of Copenhagen studied Amager island in Copenhagen to develop ways to assess *recognition justice*, which acknowledges that diverse perspectives and interests can sometimes be overlooked in the participatory planning process. The research team surveyed the island's residents about whether they felt recognized and heard in the planning of green spaces and what opportunities and constraints they perceived in the process. They also mapped the distances between partici-pants' neighborhoods and open spaces on the island. They found differences in how people of different ages and cultures use green spaces, with a marked gen-der difference between men and women; women were less likely to participate in planning processes and more likely to use green spaces closer to home. Providing green spaces, in other words, does not mean that everyone feels the same way about them or as safe in them. As we plan outdoor places for people, we need to start with people.

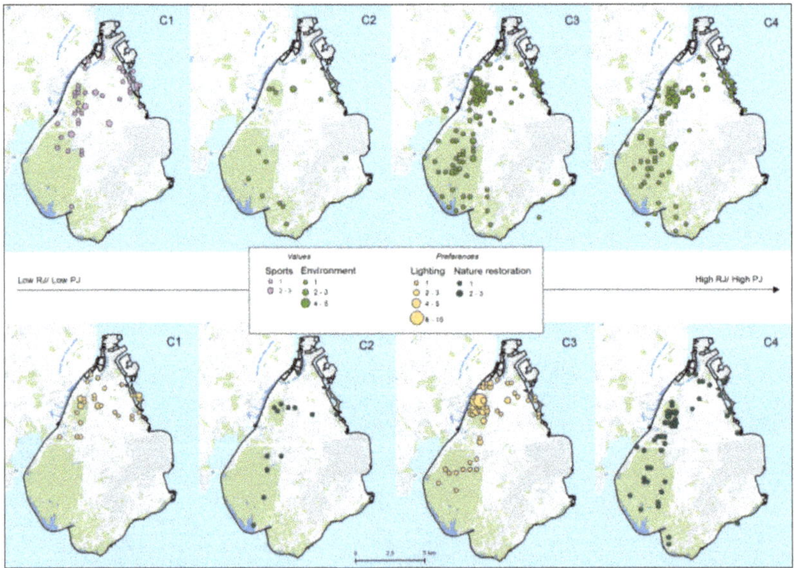

These maps of Amager island in Copenhagen, Denmark, show the preference differences among residents for sports, lighting, and access to and the preservation of the natural **environment**. *Courtesy of Silviya Korpilo, Roope Oskari Kaaronen, Anton Stahl Olafsson, Christopher Mark Raymond.*

Mapping urban plant life

DISCIPLINE Environmental Science   **INDUSTRY** Education

ORGANIZATION European Research Council   **LOCATION** Dublin, Ireland

We typically think of cities as places hostile to plants, except in the parks and other green spaces that we set aside for them. But plant life exists everywhere in cities, and this app makes that point. Developed with funding from the European Research Council and with the help of Spotteron Citizen Science and the University of Dublin, the NovelEco app lets users record and upload data about the plants they encounter in their daily lives, especially plants growing in abandoned or underused places. This volunteered geographic and botanical information helps build our understanding of feral landscapes and novel ecosystems that often get overlooked by university researchers, making such citizen-science mapping exercises invaluable. NovelEco also has a goal of increasing public awareness of rewilding processes and increasing our understanding of people's perceptions of nature in the city. The app maps the adaptations of both people and plants.

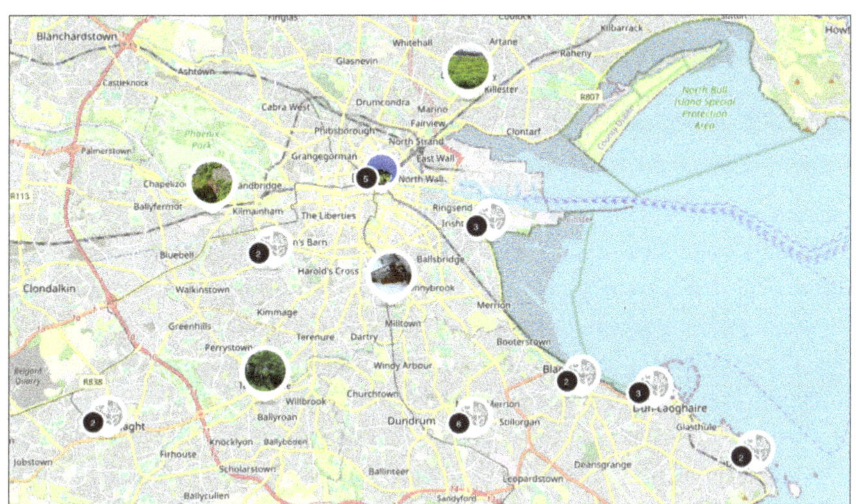

This map of Dublin, Ireland, shows where people have used the NovelEco app to identify, record, and upload data about plants they encounter in the city. *Courtesy of Spotteron Citizen Science, NovelEco, University of Dublin.*

Greening vacant urban land

DISCIPLINE Urban Forestry

INDUSTRY Sustainable Development

ORGANIZATION Temple University

LOCATION Philadelphia, Pennsylvania, USA

Philadelphia has almost 40,000 vacant lots. Since 1996, its Land Care Program has focused on removing trash and installing topsoil, planting trees, building wood fences, and cutting the grass on many of those lots. This study by two Temple University researchers looked at what effect these greening efforts have on the values of adjacent properties. The researchers found a significant positive effect in neighborhoods where the population was diverse and property values stable but a much less significant effect in neighborhoods that were either depopulating or gentrifying, suggesting that other factors affecting property values can exceed any positive benefit from the greening program. Still, the improvement of property values—as well as the psychological improvement in people's perceptions of their neighborhoods—can make such programs well worth the investment.

Achieving the 20-minute city

DISCIPLINE Public Health

INDUSTRY Transportation

ORGANIZATION Arizona State University

LOCATION Tempe, Arizona, USA

Access to various destinations in a city by multiple modes of transportation has become a major focus of planners, who often use geospatial tools to analyze that accessibility. Researchers at Arizona State University assessed Tempe, Arizona, in terms of people's access, in no more than 20 minutes, by bike, transit, and walking, to nonwork destinations, such as parks, stores, schools, and other services. Although the transit system in a sprawling, auto-oriented place such as Tempe excluded many areas of the city from a 20-minute access by bus or light rail, the study showed that residents can reach the entire city by all-roads or low-stress bike routes and by all-roads or sidewalk-only pedestrian routes. The research team hoped to further study the lived experiences of bicyclists and pedestrians in traversing the city, but their maps revealed that even a relatively low-density city such as Tempe is relatively accessible by bike and walking, at least for those with the stamina to do so in the summer heat.

Monitoring air quality

DISCIPLINE Public Health

ORGANIZATION University College London

INDUSTRY Health and Human Services

LOCATION London, England, United Kingdom

Gathering valid environmental data can be a challenge, especially when dealing with nonpoint-source pollution and air quality across a broad area. Mapping for Change, a subsidiary company of University College London, uses maps to empower people, organizations, and communities to improve the places they occupy. An example of how this works is the *Air Quality Monitoring* map they created for residents to measure nitrogen dioxide levels in their communities. The map interface allows people to locate where and when they took the reading and how far the number lies above or below the limits that the European Union has placed on NO_2 levels. This is just one of many community mapping projects undertaken by Mapping for Change, including one that maps odor pollution across Europe. The group's varied work can be summarized as mapping the change in how we know and see the world, one map at a time.

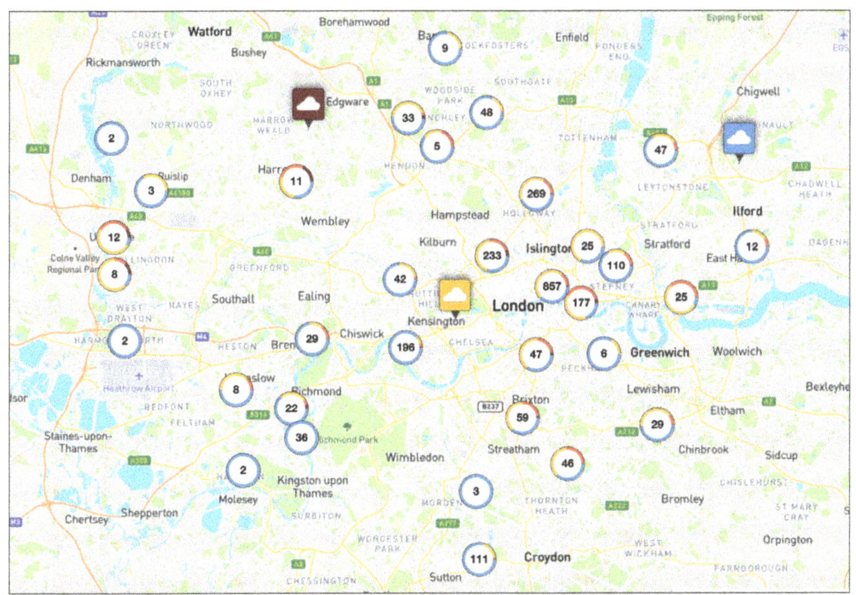

This map of London, England, shows where people have taken air quality measures and how those measures rank in terms of being above or below EU limits. *Courtesy of Mapping for Change C.I.C., University College London.*

Coordinating economic and ecological development

DISCIPLINE Geodesign

INDUSTRY Sustainable Development

ORGANIZATION Zhejiang University

LOCATION Huzhou, Zhejiang, China

The tension between economic development and environmental protection remains one of the greatest challenges around the world. This is especially true in China, which has developed faster than almost any other nation, evident in cities such as Huzhou City in Zhejiang, China. A research team from Zhejiang University and representatives from the city participated in a geodesign project that identified strategies to curb the impact of development on the local environment. These strategies included renovating rural buildings to reduce their use of coal-fired heating, encouraging urban residents to use public transit and electric vehicles to limit pollution, constructing municipal buildings and transforming agricultural practices in low-carbon ways, and restoring forests and waterways to enhance carbon-sink capacity. The map of the final design shows areas of forest restoration, agricultural transformation, and transportation and development improvements—a map of a promising future.

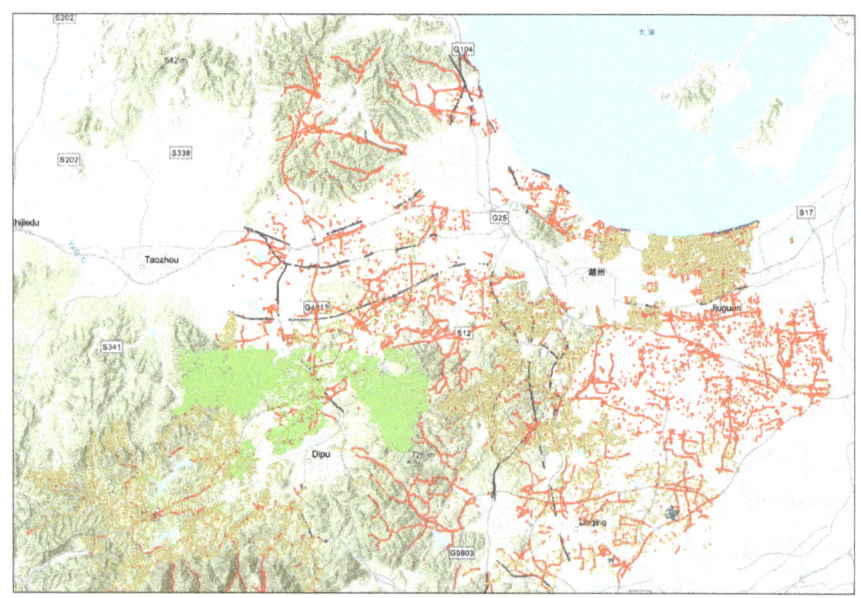

Huzhou City in China has prioritized reforestation, agricultural reform, and transformation and construction improvements to reduce its climate impacts. *Courtesy of Lu Huang and Yuchen Lou, Zhejiang University.*

Relinking city and landscape

DISCIPLINE Geodesign

INDUSTRY Sustainable Development

ORGANIZATION University of Virginia

LOCATION Winneba, Ghana

Coastal areas play a key role in creating habitat and sequestering carbon, but they are also vulnerable to industrial development and urban expansion. The slider map in this project shows the coastal city of Winneba, Ghana, and the Muni Lagoon in the Greater Accra region, indicating the current condition and the expected expansion of the city and flooding of the lagoon in 2050 if nothing changes. Based on that prospect, the research team at the University of Virginia and University of Southern California worked with students and stakeholders from the city engaged in a geodesign process, using GIS to evaluate possible future scenarios to accommodate urban growth in ways that protect the lagoon and minimize damage from coastal flooding. Out of that process came a series of strategies, including (a) densifying new development to reduce sprawl, (b) creating an urban growth boundary and core conservation area, (c) limiting the expansion of farmland and promoting alternative farming techniques, and (d) building green infrastructure and wastewater stabilization ponds to treat wastewater and runoff. Mapping the future is one of the best ways to ensure that a community will *have* one.

Note

1. Mohamed, A. A. 2025. "Mapping the Landscape of Public Participation GIS Using Natural Language Processing." *Annals of the American Association of Geographers* 115 (8): 1820–1842. https://doi.org/10.1080/24694452.2025.2511944.

References

Achieving the 20-minute city: https://gis.asu.edu/project/20-minute-city

Anticipating COVID-19 risks: https://storymaps.arcgis.com/stories /b390728f6d6f43c8bfcddf0b9e4dbbc4

Arizona State University: https://www.asu.edu/

Cagliari University: https://en.unica.it

Coordinating economic and ecological development: https://storymaps.arcgis.com/stories /aac733a1e3714c21a5359b661a508fa5

Decarbonizing Sardinia: https://b52a2f9e-824b-440e-8d87-00e7c850671a.filesusr.com/ugd /f24d78_a7f5435564624cb4a5ab9cac654c0f98.pdf

European Citizen Science Association (ECSA): https://citizenscience.eu

European Research Council: https://erc.europa.eu/homepage

Finding GIS programs in the United States: https://www.geotechcenter.org/geotech-center-national-program-locator1.html

GeoTech Center: https://www.geotechcenter.org

Going from global to national: https://mediaspace.esri.com/media/t/1_oklofdy0
Greening vacant urban land: https://journals.sagepub.com/doi/10.1068/a4595
Harvard Graduate School of Design: https://www.gsd.harvard.edu
Highlighting water scarcity in South Africa: https://www.geos.ed.ac.uk/~gisteac/gis_book_
    abridged/files/ch65.pdf
Identifying planners: https://www.tandfonline.com/doi/figure/10.1080
    /24694452.2025.2511944
League of Women Voters of Washington: https://www.lwvwa.org
Locating citizen scientists in Europe: https://citizenscience.eu/map/
Mapping the news: https://storymaps.arcgis.com/stories
    /d4a1c0037d5f41bc842f819f5b50ef37
Mapping urban plant life: https://citizenscience.eu/project/609
Mapping world ecosystems: https://geoc.umn.edu/mapping-future-collaboration-look-back-
    geocommunities-workshop
Minnesota Design Center: https://design.umn.edu/minnesota-design-center
Monitoring air quality: https://communitymaps.org.uk/project/air-quality-monitoring
National Geographic Society: https://www.nationalgeographic.org/society/
Planning for climate change in England: https://storymaps.arcgis.com/stories
    /ea8c4691fb2448238c9cac880c19f657
Planning for recognition justice: https://www.sciencedirect.com/science/article/pii
    /S0143622822001655#fig3
Planning for tsunami evacuation: https://mil.wa.gov/asset/5ba41ffb35f02
Predicting regional water demand: https://iiseagrant.org/publications/regional-water-
    demand-forecast-for-northeastern-illinois-2020-2050/
Protecting ocean biodiversity: https://umn.maps.arcgis.com/apps/mapviewer/index.html?
    webmap=922362ebbd8448ecb44335d29997aeb6
Providing election information in the United States: https://election.lab.ufl.edu/maps/
Recognizing Indigenous communities: https://mila.ss.ucla.edu/story-maps
Recovering the Indigenous landscape: https://storymaps.arcgis.com/stories
    /fe0787354a79469d99a86c2335f2ab57
Redistricting statewide elections: https://www.lwvwa.org/maps
Reducing climate impacts: https://storymaps.arcgis.com/stories
    /55b403413bc64a51a4f9bb98f7d970a2
Relinking city and landscape: https://b52a2f9e-824b-440e-8d87-00e7c850671a.filesusr.com
    /ugd/f24d78_b644bce993d04f9fa688439d539642c1.pdf
Revealing redlining: https://univredlands.maps.arcgis.com/home/item.html?id=
    6f3c336e6be345ddac8f09744bcff93e
Showing landscape change: https://clear3.uconn.edu/viewers/ctstory/
Simon Fraser University: https://www.sfu.ca
Temple University: https://www.temple.edu
Tracking opioid use: https://myutk.maps.arcgis.com/apps/dashboards
    /b3aea29bef1a4e04946e9ab8a3b7c80f
Transforming Puerto Rico: https://storymaps.arcgis.com/stories
    /4c613890e04848e797482fda18731757
UCLA: https://www.ucla.edu
Uncovering university-driven displacement: https://storymaps.arcgis.com/collections
    /dbd31671bcf84f57903ebe058537c497?item=6
University College London: https://communitymaps.org.uk

University of Connecticut CLEAR: https://clear.uconn.edu
University of Edinburgh: https://www.ed.ac.uk
University of Florida Election Lab: https://election.lab.ufl.edu
University of Helsinki: https://www.helsinki.fi
University of Illinois Extension: https://iiseagrant.org
University of Lisbon: https://www.ulisboa.pt
University of Lodz: https://www.uni.lodz.pl/en/
University of Manchester: https://www.manchester.ac.uk
University of Minnesota GeoCommons/GeoCommunities: https://geoc.umn.edu/
University of Redlands: https://www.redlands.edu
University of Tennessee Knoxville: https://www.utk.edu
University of Vermont: https://www-igcollab.hub.arcgis.com
University of Virginia: https://www.virginia.edu
University of Wisconsin—Milwaukee: https://uwm.edu/libraries/agsl/
Washington Military Department/WA Emergency Management: https://mil.wa.gov
Zhejiang University: https://www.zju.edu.cn

Credits and contributions

ProtectedSeas Navigator Map of Marine Conservation Regulations. ArcGIS Living Atlas of the World, ProtectedSeas®. Accessed December 23, 2025. https://protectedseas.maps.arcgis.com/home/item.html?id=922362ebbd8448ecb44335d29997aeb6.

Hurley, Stephanie, Hannah Turner, and Gillian Galford. "Winooski River Watershed, Vermont, USA, Geodesign StoryMap" for International Geodesign Collaborative.

February 2024. https://storymaps.arcgis.com/stories/55b403413bc64a51a4f9bb98f7d970a2.

Tilton, Jennifer, Lisa Benvenuti, and Tessa VanRy. People's History Race Dot Density Map 1900-2020. A People's History of the Inland Empire Census Project Using IPUMS Ancestry Full Count and IPUMS USA: Version 12.0. Program in Race and Ethnic Studies and Institute for Geospatial Impact. University of Redlands and UCR Public History. 2025.

App and Marker: SPOTTERON Citizen Science. www.spotteron.net. Map: (c) OpenStreetMap contributors. www.openstreetmap.org.

CHAPTER 6

GIS careers

Introduction

Previous chapters have looked at how colleges and universities around the world use GIS for academic research, classroom teaching, facilities management, and community engagement in a variety of ways. This chapter focuses on something slightly different but no less relevant to higher education: how employers use GIS in their work and what jobs await students with GIS skills after they graduate. Although a college degree has many benefits beyond career preparation, the latter certainly factors into what colleges need to teach and students need to learn. And as this chapter aims to show, preparation for almost any career now needs to include some knowledge of GIS.

Graduates will need GIS for various purposes, depending on the line of work and types of careers they pursue. In the part of the economy that produces and delivers goods, the locational and navigational power of GIS has become an essential feature of businesses that source materials, route products, and deliver them to customers, either in stores or directly to their doors. The same need exists on the services side of the economy, where the delivery of information—when and where people need it—has become equally dependent on GIS as a knowledge delivery and data visualization tool. The Science of Where now applies to almost every aspect of the modern economy.

This chapter focuses on the industries that make up the major use cases of GIS in the workplace. For each major industry, we will offer one or two examples, with the goal of giving students interested in becoming GIS professionals an overview of the possible career paths they might follow after graduation. All these industries employ hundreds of thousands to millions of people just in the United States alone, and the demand for people with GIS skills is only expected to grow as the geographic approach has become central to the modern economy.

In addition to preparing students for a variety of careers, GIS can help them map their paths to these careers. Many colleges and universities use data-informed, interactive maps to help students and recent graduates find employment and enable career-services staff to know who has gone where after graduation. Northwestern University's career advancement office, for example, has a website that shows where graduates are working, with data on the percentage of graduates who are employed, searching for work, or still in school or the military. The website also includes data on the industries and companies where graduates are working, as well as their graduate programs. Maps can help people find their way in life.

The GIS workplace

Architecture, engineering, and construction

The GIS-related jobs in the architecture, engineering, and construction (AEC) industry include landscape architecture and urban design, civil and environmental engineering, construction management and materials logistics, and now—with the integration of GIS and CAD—interior design and facilities management.

Business

Constituting almost half the workforce in many countries, private-sector organizations use GIS in the insurance and financial services industries, product logistics and distribution, mining and manufacturing, and real estate and retail activities. The on-demand delivery economy, now a major aspect of many businesses, depends on the locational capacity of GPS and GIS.

Conservation

Although one of the smallest industries in terms of the number of people working in the field, conservation remains a major user of GIS for land management and landscape preservation activities as well as community-based activism related to the environment. GIS does not just enable this work; it also reveals the reasons that this work is so important in the face of declining biodiversity.

Defense and intelligence

The defense and intelligence industries remain one of the largest components of government budgets in many countries, and GIS plays a central role because defense is fundamentally a spatial activity—a matter of defending territory—as is defense-related intelligence, in its effort to understand what others—friends and enemies—are doing where.

Education

The sector size of primary, secondary, and higher education shows the importance that many countries place on education. And GIS—as this book shows—has become an important way to teach students about the world, which is inherently spatial, and address the fact that visual forms of communication, such as maps and aerial photography, remain powerful tools in helping students learn.

Energy utilities

Although energy itself may be invisible, the infrastructure that generates and delivers it is certainly not, and GIS has become an essential tool for electric and gas utilities to locate their facilities, route their distribution lines, and manage the assets that they must maintain for the rest of the modern economy to function.

Health and human services

Human health depends on the environments in which we live and work, and GIS plays a critical role in locating and addressing the factors that affect public health and determine access to health care. GIS also enables the delivery of many other human services that people depend on to achieve their fullest potential, such as getting food, medicine, or emergency aid to where it's most needed.

National government

This part of the public sector has grown more slowly in many countries, but national governments use GIS in almost all aspects of their operations, including managing civilian and military aviation, election integrity, humanitarian assistance, historical sites, national parks, official statistics, national mapping, and maritime monitoring.

Natural resources

This industry has perhaps the largest share of land area, although its workforce remains among the smallest of the major industries, in part because of the industrialization of agriculture, forestry, mining, drilling, and renewable energy. GIS has helped those industries become more efficient and reduce their negative impact on the environment.

Nonprofits and NGOs

The nonprofit sector—with its many, relatively small organizations—has become a major employer and an important user of GIS to locate supported communities and activities. GIS also serves as a way to communicate the scope of nonprofit and NGO missions and measure their impact.

Petroleum

The complexity of locating oil and gas reserves and then extracting, refining, and distributing these products where they're needed requires spatial knowledge that only GIS can provide. The petroleum industry also has a tremendous amount of infrastructure—pipelines, refineries, storage facilities, ports—that must be managed, using GIS as a vital tool.

Public safety

Public safety is location specific; ensuring public safety requires GIS to anticipate and locate crime, plan for and respond to emergencies, and identify where fire, rescue, and emergency medical services are needed. GIS also enables responses to national security threats, humanitarian assistance needs, and wildfire events, among the many other risks to safety.

Science

Much of this workforce is based in colleges and universities or in government-sponsored research labs, and an increasing amount of scientific activity involves GIS, especially in spatially oriented fields such as weather and climate science, oceanography, earth sciences, and social sciences, including geography and GIS itself.

State and local government

Often larger than the national government, local and regional governments use GIS for a range of activities, including economic development, environmental and natural resources, health and human services, housing and serving the unhoused, land records, public works and engineering, and urban and community planning.

Sustainable development

This sector remains the hardest to quantify in terms of the number of workers, depending on how we define it. However, it's also one of the fastest growing parts of many economies, with some of the greatest growth in clean-energy jobs. The entire industry uses GIS to locate where sustainable development can have the greatest impact for the greatest number of people.

Telecommunications

Like other infrastructure-related industries, telecommunications may not have a large workforce, but we all depend on it to lead our connected lives. Telecommunications companies use GIS to plan and engineer, operate and maintain, and sell and market their networks, which stretch across the globe and in space in the form of telecommunications satellites and undersea cables.

Transportation

The movement of people and goods persists as one of the most important aspects of the global economy, and GIS facilitates it all, whether it involves the movement of aircraft at airports, ships at ports, trains at stations, buses at bus stops, or commercial and passenger vehicles on their way to their destinations.

Water

Cleaning, conserving, and carrying water from its source to its place of consumption require GIS along the entire supply chain. Water utilities have developed sophisticated ways of mapping the flow of water, locating leaks or other disruptions in the system, and identifying the best ways to preserve this vital resource.

Architecture, engineering, and construction examples

Architecture

DISCIPLINE Arts

ORGANIZATION Marcus Neustetter

INDUSTRY Architecture

LOCATION Johannesburg, Gauteng, South Africa

Architects use maps to determine their building sites and understand the contexts in which their work stands. The widespread use of GIS among architects, however, rarely gets visualized as directly as in this apartment building in Johannesburg, where the South African artist Marcus Neustetter has applied a GIS-based map of its neighborhood on the facade.[1] Neustetter has long been interested in the impact of pixelation on our experience of computer images and on how changes in the scale of an image can alter our understanding of it. This work reminds us that the idea of buildings having stories about them comes from a medieval tradition in Germanic countries of painting scenes from the lives of their owners on the fronts of their houses. Neustetter's installation also reminds us that a floor plan is itself a kind of map, one that we navigate constantly in our daily lives as much as we do the larger landscape. In the end, maps tell stories and it seems appropriate—if unconventional—to place a map on the stories that make up a building.

The relationship of a building to its location is highlighted in this map installation of the neighborhood map on a building's facade. *Courtesy of Marcus Neustetter.*

Landscape architecture

DISCIPLINE Arts

ORGANIZATION Esri

INDUSTRY Landscape Architecture

LOCATION Netherlands

GIS, in some ways, grew out of the field of landscape architecture and the work of landscape architects Ian McHarg, who envisioned landscapes as overlapping and interacting layers, and Jack Dangermond, who, as Esri's cofounder and president, helped bring that idea into the digital world. As a result, landscape architecture has become a richer field, not just with the creation of beautiful outdoor spaces but also one grounded in a range of environmental and social sciences. This story map[2] of open spaces in Rotterdam speaks to that diversity. The 3D image of this park that's part of the story map, indicating belowground utilities and the places and plant material on the surface, shows how GIS helps landscape architects see how their designs affect the hydrology and infrastructure of the place and how people might use and experience parks. In this case, GIS also lets landscape architects assess the impact that an adjacent building will have on the park and adjust their plans accordingly, showing how GIS is a powerful design tool as well as an analytic one.

GIS allows designers to create parks and public spaces that account for the layers of infrastructure above and below ground. *Courtesy of Esri.*

Business examples

Consumer behavior

DISCIPLINE Business 　**INDUSTRY** Business
ORGANIZATION GIS Geography 　**LOCATION** Global

Predicting consumer behavior remains a key concern in the retail industry. The Huff gravity model offers one approach, looking at the probability of a customer visiting a store based on its attractiveness and distance to other similar sites. A retailer employing this model will look at census data and assess the size of each store and its distance to each census tract to determine the store's attractiveness. To measure the probability of each store's market share, the model has the retailer divide the attractiveness of each store by the total attractiveness of all the stores. The map shows what the model can produce. The highlighted store will probably capture the greatest market share, whereas the others shown will each have an equal chance that customers might go to any of them. The maps produced by the Huff gravity model are invaluable in determining which stores consumers will most likely visit and which ones, unlike the goods in them, are up for grabs.

Real estate

DISCIPLINE Business 　**INDUSTRY** Business
ORGANIZATION JLL 　**LOCATION** Washington, USA

In the real estate industry, it's location, location, location. And GIS has proven invaluable in helping real estate professionals identify locations for new development, infrastructure upgrades, or neighborhood revitalization. JLL, the world's second-largest real estate brokerage by revenue, uses GIS to help companies identify sites for possible facilities, conduct demographic comparisons, and track population trends. The brokerage also uses interactive maps to analyze access of industrial sites to infrastructure, such as rail lines, highway entrances, and shipping facilities, using its own curated source of maps and data to speed the process. GIS helps JLL clients visualize information in ways that everyone can easily understand.

Conservation examples

Biodiversity

DISCIPLINE Biology

ORGANIZATION Esri

INDUSTRY Conservation

LOCATION USA

Biologists have an enormous scope of study, with all of life on earth within their purview, and maps can help identify the boundaries of their research. This map of North America's grasslands offers one such example, focusing on an especially threatened ecosystem. This map indicates where 62 percent of these grasslands have been converted into farms or development and where as much as 93 percent of the remainder faces eradication by the same forces. Other maps document naturalists' observations of plants, birds, and insects as well as the locations of one of the grasslands' keystone species—the prairie dog—whose "towns" often extend underground over vast territories. Prairie dog towns are a possible model for how humankind might occupy this ecosystem in the future—largely underground, with minimal disturbance of the sea of grass above.

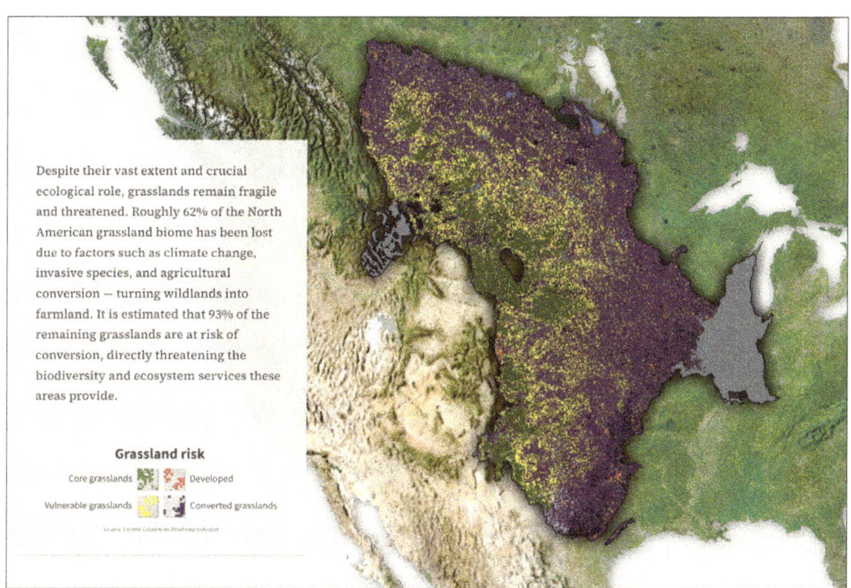

This map of North America's grasslands shows how much has been lost and how much more is threatened in the future. *Courtesy of Esri.*

Ecology

DISCIPLINE Conservation

ORGANIZATION Refractions Research

INDUSTRY Land Development

LOCATION Vancouver, British Columbia, Canada

Ecologists excel at synthesizing a lot of data into a comprehensible form to understand the biodiversity of a region. As an example, the nonprofit Biodiversity BC asked the GIS consulting firm Refractions Research to consolidate some 30 layers of data into a user-friendly, open-source tool that allowed people to understand the biodiversity of British Columbia in new ways. Users can prompt the tool to look for related features, such as slope, land cover, and climate, to locate particular ecosystems and create summaries of how much of a particular ecosystem exists in the province. The tool also lets users view the data layers in the system, such as the location of the province's ecological drainage areas. Biodiversity is inherently spatial and inevitably complex, and we have only recently developed tools that are geospatially complex enough to understand it and, as in the case of British Columbia, protect it.

Defense and intelligence examples

Defense

DISCIPLINE Aviation

ORGANIZATION Esri

INDUSTRY Defense

LOCATION Global

The defense industry is one of the largest users of GIS in many countries, in part because geospatial tools allow military personnel to combine a lot of data in real time and three dimensions. The ArcGIS 3D Analyst™ extension, for example, allows pilots to avoid routes where they might be targeted, requiring them to understand vertical and horizontal distances. This map simulates the location of three antiaircraft guns, with different ranges depending on the presence or absence of radar. This lets planes avoid the orange and green zones, although the latter, requiring radar assistance, presents less of a hazard and allows pilots to take countermeasures. These threat domes, in other words, enable pilots to calculate the risk of various flight paths in real time and in a readily understood way, mapping their way to safety.

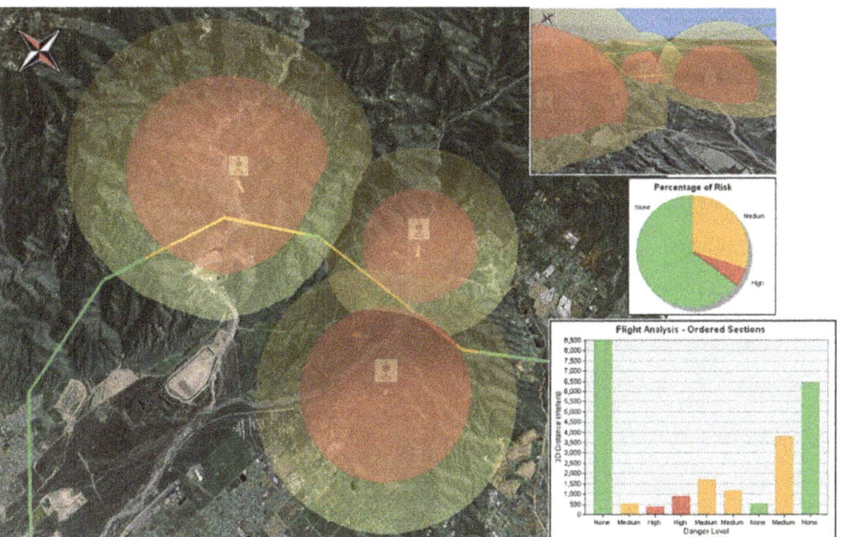

Pilots need to think in three dimensions, especially as they face threat domes created by antiaircraft guns. *Courtesy of Esri.*

Intelligence

DISCIPLINE Aviation   **INDUSTRY** Aviation

ORGANIZATION Jason Davies   **LOCATION** Global

The aviation industry often has a different perspective on the planet, not only because of pilots' aerial viewpoints but also because of their need, especially during a flight emergency, to determine which airports are the closest. This Voronoi diagram of world airports shows, from any point on the planet, how a pilot can find the nearest place to land. Such diagrams identify seed points—in this case, the location of airports—and determine an area based on all the points closer to one seed point than any other. This is most accurate when shown on a 3D sphere to eliminate the distortion of two-dimensional maps. The airport Voronoi diagram also indicates, in places such as Europe or India, where other forms of transportation, such as trains, may be faster and more efficient than flying, considering the density of population and proximity of airports. And the diagram ensures that when we do fly, no one is flying blind.

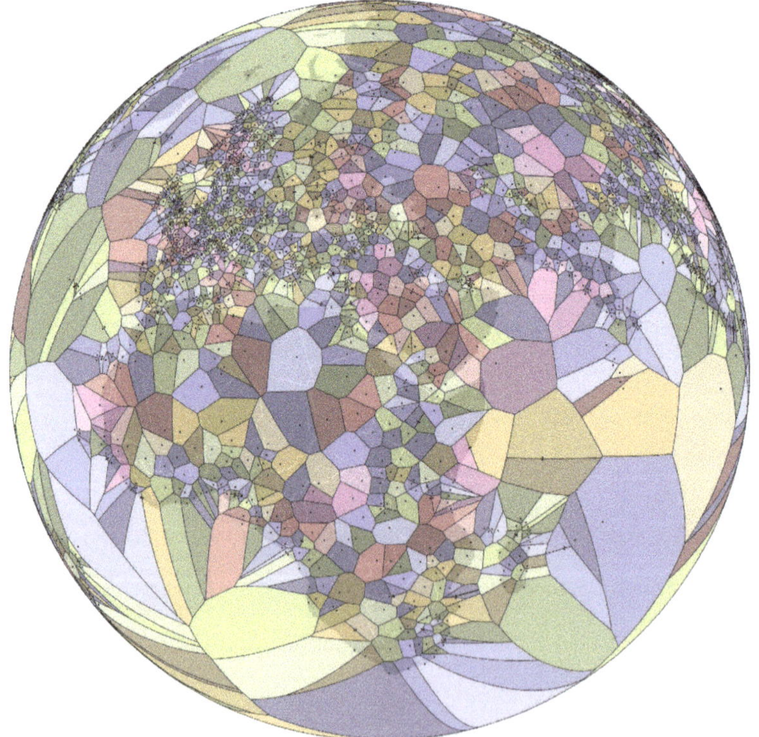

For pilots in an emergency, knowing the location of the nearest airport can be a matter of life or death. *Courtesy of Jason Davies.*

Education example

<table>
<tr><td colspan="2" style="background:#c17a45;color:#fff">Equity</td></tr>
<tr><td>DISCIPLINE Public Policy</td><td>INDUSTRY Education</td></tr>
<tr><td>ORGANIZATION Austin ISD</td><td>LOCATION Texas, USA</td></tr>
</table>

One of the core values of public education is to provide the best possible education to every student, regardless of background or ability. Maps can sometimes show the gaps between that goal and the reality on the ground, even in places as affluent as Austin, Texas. This study of Austin's public schools highlights the overlap among the number of underserved students, the location of socially vulnerable neighborhoods, and the number of facilities most in need of repair, showing the extent to which schools reflect their communities, for better or worse. The Austin school system has made this information public and used it as the basis for conversations with diverse communities about what the maps show. Whether or not that overlap was by design, design can help educational leaders come up with strategies that address the needs of those in their communities who have been underserved for too long.

Energy utilities example

<table>
<tr><td colspan="2" style="background:#c17a45;color:#fff">Electric grid</td></tr>
<tr><td>DISCIPLINE Electrical Engineering</td><td>INDUSTRY Energy Utilities</td></tr>
<tr><td>ORGANIZATION Oregon DLDC</td><td>LOCATION Oregon, USA</td></tr>
</table>

Managing assets in the electric utilities sector remains a major challenge, considering the size and complexity of the electric grid, especially in the United States. The US electric utility GIS map locates the major electric utility lines and allows users to search by an address to find the service lines that deliver power to that place. The map indicates the concentration of utility lines in large cities and more populous states and illustrates the interconnectedness of the US electric grid. It can lead to greater resilience in the system through the redundancy that characterizes nature's networks—more local production and consumption of energy and less dependence on the extensive infrastructure evident in the map.

Health and human services examples

Dentistry

DISCIPLINE Public Health

INDUSTRY Health and Human Services

ORGANIZATION Iowa Dept. of Public Health

LOCATION Iowa, USA

Dental health depends on access to dental care, levels of water fluoridation, and other factors. GIS can show the relationship between these factors as this Iowa community water fluoridation map indicates. The map marks the location of unfluoridated water systems and fluoridated ones. Some counties have a higher percentage of tooth decay among youth, and others have the greatest shortage of dentists per capita. Such data visualizations reflect the complexities of human health—higher tooth decay exists in some counties with fluoridated water and lower levels of tooth decay are found in some counties without. Overall, the map shows that fluoride can reduce tooth decay in young people, as research has shown since the early 1900s. It's enough to make a dentist smile.

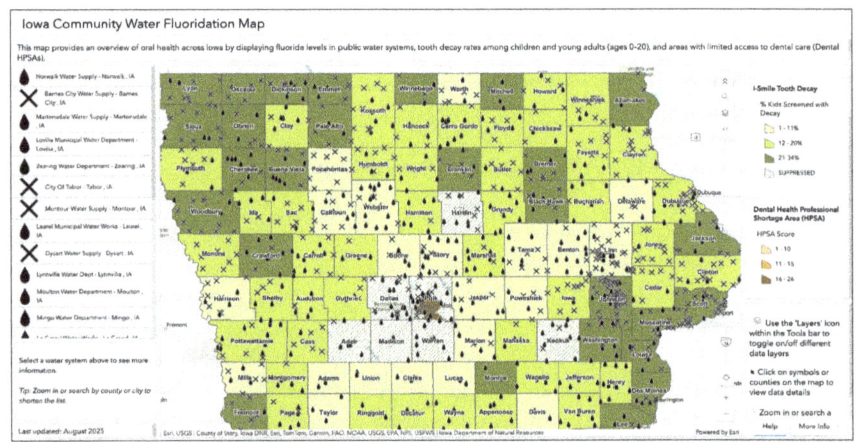

The complex relationship between fluoridation and tooth decay is shown in this publicly accessible map. *Courtesy of Iowa HHS, Oral Health Section, Division of Public Health.*

Public health

DISCIPLINE Public Health

INDUSTRY Health and Human Services

ORGANIZATION World Health Organization

LOCATION Global

Public health as a discipline arose in the 19th century in response to the needs of populations around the world and the spread of contagious diseases as transportation made it easier for people to travel from one part of the world to another. Today, the World Health Organization leads that work globally; its tracking app of COVID-19 transmission shows how much the mapping of disease occurrence is a crucial part of bringing epidemics and pandemics under control. The app maps the number and location of COVID cases over the last week, last month, and cumulatively since the start of the pandemic, with each dot indicating that country's data on the disease. A major accomplishment of public health lies not just in its sharing of health data and coordination of emergency response, but in its commitment to making the information publicly accessible in easily understood apps such as this.

National government examples

Economics

DISCIPLINE Business

INDUSTRY NGOs

ORGANIZATION World Bank

LOCATION Global

Banking may not seem like a spatial activity, in part because of the mobility of money and movement of capital, but banks and their economists also serve particular places and identify needs and measure risks based on locations, making economists especially dependent on maps. The World Bank, for example, uses maps as the basis for providing its data about the economies of every country. Its website has a map showing the relative income, overall population, and gross domestic product of each nation. It then links to a profile page that shows in detail a range of social, economic, and environmental indicators, along with lists of development projects, human capital indices, and economic and climate change forecasts. Maps, it seems, are something that even bankers can take to the bank.

Demographics

DISCIPLINE Geography   **INDUSTRY** Education

ORGANIZATION Esri   **LOCATION** USA

Physical and human geography map two different things, and this becomes apparent when looking at the geographic center and the population center of a nation, evident in a map using Esri's Mean Center and Spatial Statistics tools. In the case of the United States, with its origins on the Eastern Seaboard, the center of its population stood approximately where Washington, DC, now stands. As the country grew to the west, so did its population center shift to the Ohio–West Virginia border in 1850, then to Indiana, Illinois, and now Missouri over the course of the 20th century. By 2005, it had moved close to the state's capital, Jefferson City, named after the US president who helped launch the westward growth of the country. That geographic center of the Lower 48 shows how the country has a more densely populated eastern third. It also highlights the interdependence and distinction between two of the most fundamental social sciences: geography and demography.

Natural resource examples

Agriculture

DISCIPLINE Agriculture

ORGANIZATION USDA NASS

INDUSTRY Natural Resources

LOCATION USA

Agriculture represents a major part of many nations' economies, and knowing what crops can grow and where has made agriculture a more efficient process, taking advantage of soil types and weather conditions in ways not possible in the past. This interactive map[3] of the United States, called *CropScape*, was developed by the US Department of Agriculture's National Agricultural Statistics Service and George Mason University's Center for Spatial Information Science and Systems. *CropScape* allows the agricultural community to zoom to locations to see the distribution of crops, with annual data layers that go back to 1997, showing average estimates of crop types. The map also allows policymakers to assess food access, predict land-cover change, and help farmers deal with weather and landscape challenges. In an era of great change in the agricultural sector and the global climate, the ability of farmers to know, precisely, what types of crops grow best and where can make all the difference between having a healthy harvest or not having one at all.

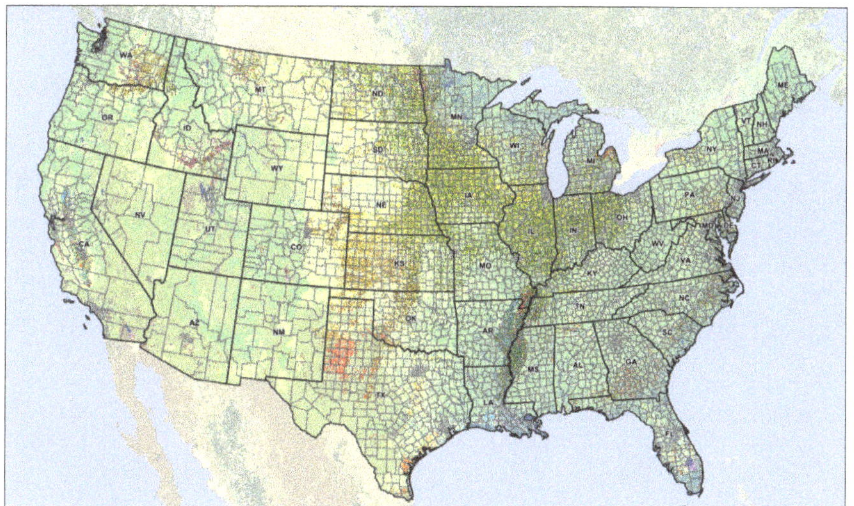

CropScape offers a county-by-county map of what gets planted where in the United States, with data layers available going back to 1997. *Courtesy of USDA.*

Agronomy

DISCIPLINE Agriculture   **INDUSTRY** Natural Resources

ORGANIZATION ISRIC   **LOCATION** Global

Agronomy and soil science deal with one of the most pervasive and unseen parts of our lives: the ground beneath our feet. Yet much of what we do above ground depends on the nature of what lies below. An example of that is Soil Grid,[4] an app that uses GIS to document the soils around the world in support of the global agricultural community, which needs to know what kind of plant materials can thrive where. Soil Grid enables users to search by soil classes as well as the chemical composition and other properties of soil anywhere in the world. It also lets us zoom to a one-kilometer scale to determine soil conditions at a local level, enhancing its usefulness to farmers and horticulturalists whose plans depend on knowing that information. It's an app grounded in what matters most about the ground.

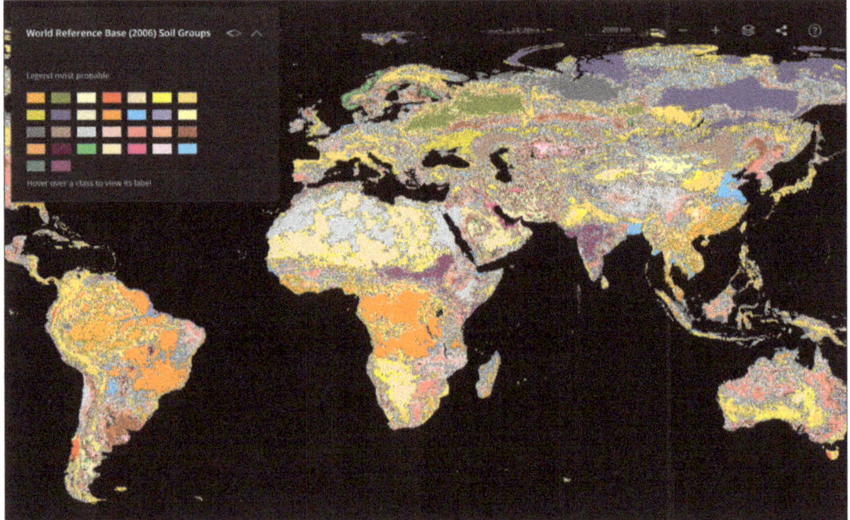

This map of soil classes around the world, based on a one-kilometer grid, tells us a lot about what we can plant where. *Courtesy of Soil Grids.*

Nonprofit and NGO examples

Climate change

DISCIPLINE Climate Science

ORGANIZATION Trubel & Co

INDUSTRY Sustainable Development

LOCATION Florida, USA

Students are often more willing to face up to the world's challenges than many of the adults who helped create those challenges in the first place. Trubel & Co. is a nonprofit in Florida focused on technology justice, giving marginalized students skills in GIS and geospatial analytics to show the climate change challenges and environmental injustices facing vulnerable low-income communities. Through their data collection and mapping efforts, students showed how hundreds of farms in southwest Florida are at risk of becoming nonarable because of rising sea levels. This will affect not just the farms impacted by saltwater inundation but also the communities of which the farms are a part, as well as the food system that depends on the agricultural products of the region.

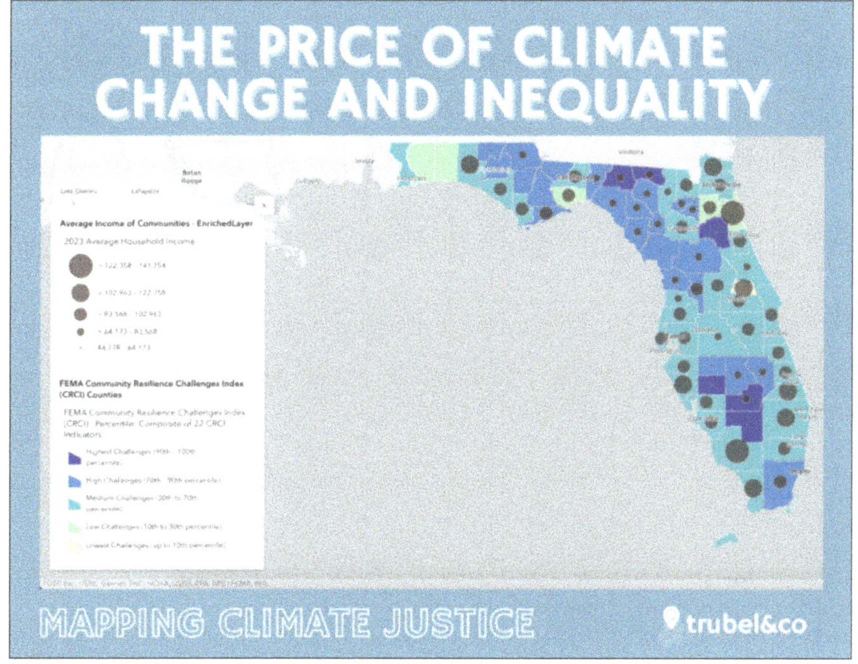

Coastal communities will suffer the most from climate change in the coming years, especially those communities without the means to move. *Courtesy of Trubel & Co.*

Digital equity

DISCIPLINE Public Health   **INDUSTRY** Education

ORGANIZATION University of Minnesota   **LOCATION** Minnesota, USA

The COVID pandemic showed that people who didn't have access to a digital device or the internet suffer by being excluded from the online economy: online learning, shopping, and working. Smart North, a nonprofit in Minnesota, responded to that need by setting up tech hubs in schools, clinics, and community centers to give underserved youth access to today's digital tools and teach them GIS as a core skill needed for the online economy. With the support of the University of Minnesota's GeoCommunities initiative, the tech hub in Essentia Health's clinic in Deer River, Minnesota, has been training Indigenous youth from the Leech Lake Band of Ojibwe in GIS. The goal is to engage youth in planning the future of their tribal lands and empower them to envision the community where they want to live, giving them sovereignty over their data so that they maintain control over that future.

Petroleum example

Pipelines

DISCIPLINE Engineering	**INDUSTRY** Petroleum
ORGANIZATION Esri	**LOCATION** USA

Delivering everything from gasoline and jet fuel to home heating oil keeps the American economy moving and American households warm in the winter. Tracking the location of the major petroleum pipelines by the US government's Energy Information Administration (EIA) is critically important to ensure the system's safety and security. The GIS map[5] shows the roughly 64,000 miles of pipelines, ranging from 12 inches to 42 inches in diameter, and the sites of fuel terminals, where trucks move product from petroleum storage tanks to gas stations and home fuel tanks. A part of ArcGIS Living Atlas of the World, the map is regularly updated by the EIA, which collects, analyzes, and disseminates impartial energy information, including its impact on the economy and environment.

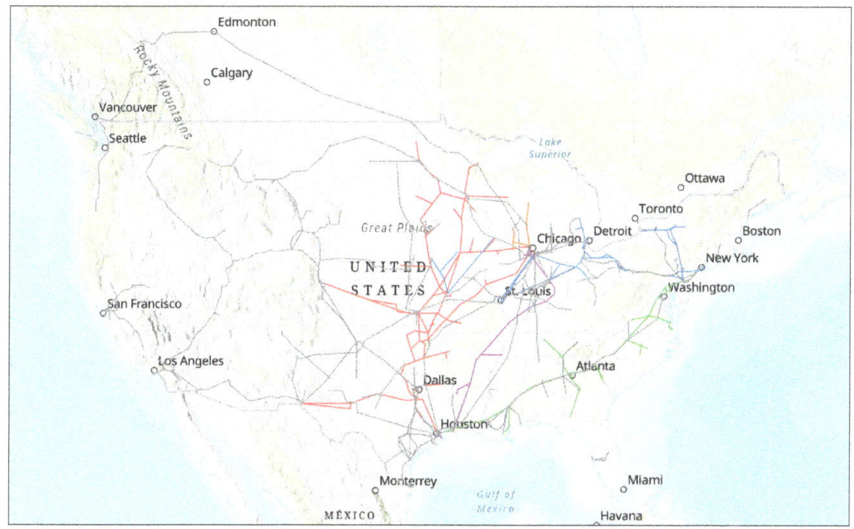

The pipelines that deliver the oil and gas that keep America running represent a dense infrastructural network across the country. *Courtesy of Esri.*

Public safety examples

Criminology

DISCIPLINE GIS   **INDUSTRY** Education

ORGANIZATION Esri   **LOCATION** Djibouti

Criminology tries to understand the nature of crime and to anticipate when and where it will occur to prevent crimes from taking place. In a map of where criminals might meet or congregate in Djibouti, crime fighters have used phone data to track the movement of criminals and align the data with temporal data, such as minimum loiter time. If the phone tracking in any location is greater than the minimum loiter time, it indicates a possible meeting of suspected criminals, enabling the police to focus their time and attention on those locations. Such pattern recognition can also allow law enforcement to anticipate meeting places and intervene in criminal activity before it occurs. Such maps show that crime is as much a spatial concern as a legal one, and GIS can use the one to monitor the other.

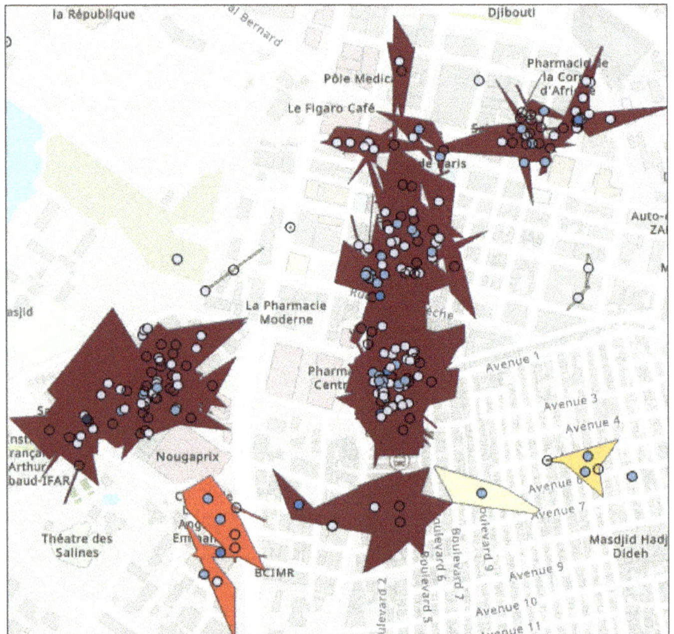

GIS has enabled crime fighters to predict where crime may occur and criminals may congregate, reducing the overall crime rate. *Courtesy of Esri.*[6]

Physical safety

DISCIPLINE Public Health   **INDUSTRY** AEC

ORGANIZATION Esri   **LOCATION** Ohio, USA

Public safety extends beyond the public realm to include private industries, such as amusement parks, which have used digital technology to immerse people in alternative realities or future environments. A 3D animation of the Millennium Force roller coaster at the Cedar Point amusement park in Ohio shows the coaster from two perspectives: that of the engineer concerned about the safety of the structure and the geometries of the coaster as well as that of the user interested in what it would feel like to ride on the coaster. The ability to use existing data to simulate the movement in a 3D scene has made GIS an exceptional tool in the creation of fly-through animations. It has also become invaluable in the entertainment industry to make its offerings as entertaining and as safe as possible.

Science example

Seismology

DISCIPLINE Seismology   **INDUSTRY** Science

ORGANIZATION Student   **LOCATION** Global

Mapping seismic activity has enormous benefits, not only for the communities most affected by volcanoes and earthquakes along tectonic plate boundaries but also for people and communities thousands of miles away, exposed to tsunamis produced by those events. What may seem like a spatially confined event threatens much of the planet in one way or another. As this study of earthquakes, volcanoes, and tsunamis describes, their greatest concentration occurs along seismic faults near coastlines or in some of the deepest parts of the ocean. However, large intercontinental areas can feel the effects of seismic activity, and almost any ocean coast can be vulnerable to the threat of tsunamis. When it comes to the consequences of plate tectonics, it isn't just a matter of where it will affect people, but also how and when.

State and local government example

Geography

DISCIPLINE Geography   **INDUSTRY** Science

ORGANIZATION what3words   **LOCATION** USA

Communications jobs take many forms, and yet almost all involve words and images. When it comes to communicating something about a place, maps offer a valuable and easily available way of doing so, but maps also depend on words to describe or annotate them. That interdependence of visual and verbal forms of communication is especially apparent in the what3words geolocation app.[7] Its developers have laid a three-meter-square grid over the entire planet and attached a three-word identifier to each of the millions of grid units, allowing users to locate any nine-square-meter place on the planet by knowing its three-word moniker. That simple idea has multiple applications, enabling emergency responders to find disaster victims, ride hailers to find carpool drivers, or repair workers to find malfunctioning equipment. And the use of words to find places suggests a new type of communication, one in which the visual and verbal merge into an equivalent and codependent form of navigation.

Attaching three words to every unit of a three-meter-grid overlaying the world allows us to find locations with much greater precision. *Courtesy of what3words.*

Sustainable development example

United Nations Sustainable Development Goals

DISCIPLINE Public Policy

INDUSTRY Sustainable Development

ORGANIZATION University of Minnesota

LOCATION USA

The UN Sustainable Development Goals (SDGs) have provided a framework that focuses the work of governments and nonprofits to improve environmental and human health and encourages interdisciplinary efforts in higher education. The University of Minnesota has a systemwide SDG initiative to support and promote transdisciplinary research, teaching, and partnerships; its College of Education and Human Development has embraced this initiative enthusiastically. The college looked at its faculty research and scholarship through its experts database and realized that its faculty has made many contributions to the SDGs, including 160 contributions to SDG 3 (Good Health and Well-Being), 74 contributions to SDG 16 (Peace, Justice, and Strong Institutions), 61 contributions to SDG 4 (Quality Education), 31 contributions to SDG 5 (Gender Equality), and 25 contributions to SDG 10 (Reduced Inequalities). The SDGs offer a way for higher education to show how its work remains relevant to the world and the world of work.

Telecommunications examples

Cables

DISCIPLINE Civil Engineering   **INDUSTRY** Telecommunications

ORGANIZATION Esri   **LOCATION** Global

Civil engineering largely emerged in the 19th century to address the infrastructure needs of civil society, as opposed to the military. The field has grown to affect large parts of the planet, including the oceans, with the many communications cables that run through the ocean depths linking continents to one another. This app, designed and developed by Esri's Applications Prototype Lab, shows the locations of the global network of submarine communication cables with a 3D map that allows users to search by the name of a cable and its landing points. Although this map is useful to the companies and countries laying such cables, the app also has public awareness value in revealing the extent of the submarine infrastructure that links coastal nations. Anyone who thinks that a country can go it alone need only look at this app to see how impossible such an idea would be.

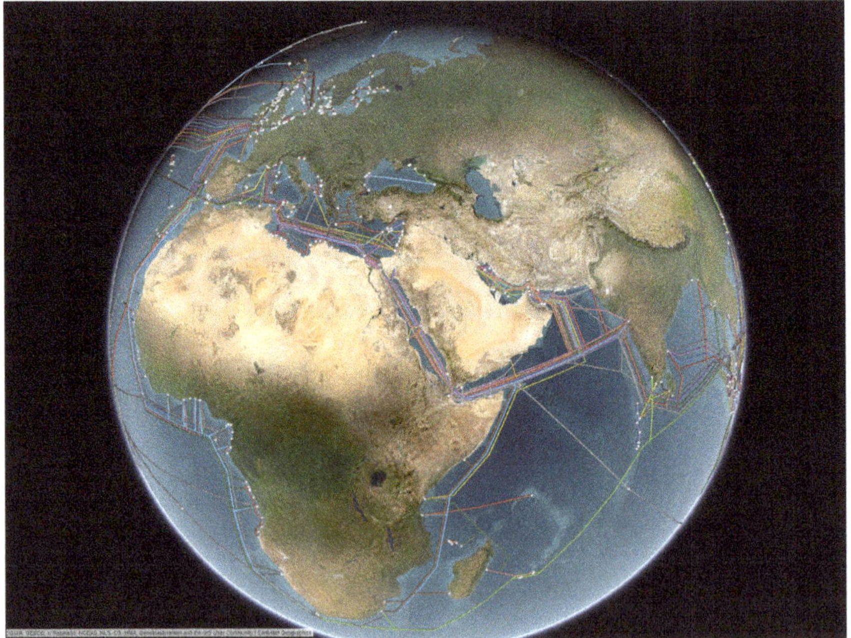

Although satellite communication has grown, so too have the number of communications cables in the oceans. *Courtesy of Esri.*

Satellites

DISCIPLINE Aerospace

ORGANIZATION Esri

INDUSTRY Science

LOCATION Space

We typically think of maps as mainly terrestrial in nature, allowing us to locate things on the earth. But outer space has its own geography and locational challenges, as shown in this map of the 18,792 satellites circling the planet. Developed by Esri's Applications Prototype Lab, the interactive app uses data from the nonprofit Space-Tracker, which shares information about satellite locations with the private and public sector in charge of satellite operations to promote the "peaceful use of space worldwide." The app allows users to search for satellites based on the countries that launched them, the type and size of equipment, the launch dates, and the nature and location of their orbits. It provides a critical tool for those who navigate the hardware that circulates above our heads and illustrates why we need more cooperation among nations about not only the location of satellites but also their number. Our terrestrial world may be crowded, but so is our celestial one.

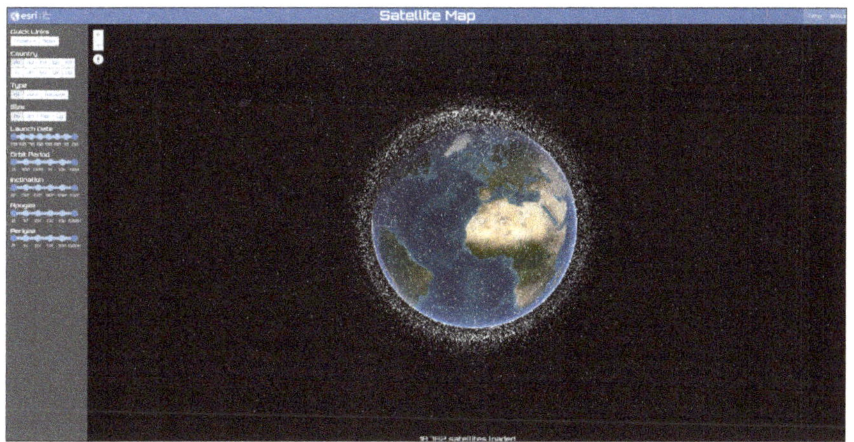

This satellite map shows the location of space technology circling the planet. The data is sortable by type, size, date, and source of each satellite launch. *Courtesy of Esri.*

Transportation examples

Noise

DISCIPLINE Science and Technology   **INDUSTRY** Transportation

ORGANIZATION US Department of   **LOCATION** Illinois, USA
Transportation

Acoustics has long been concerned with indoor spaces, such as concert halls and sports venues, but acousticians have increasingly focused on the outdoors and the challenge of mapping the less predictable nature of noise in the environment. A good example is the *National Transportation Noise Map*. Produced by the US Department of Transportation's Bureau of Transportation Statistics, the map[8] can help decision-makers, researchers, and the public visualize noise patterns related to airports and major railyards, as in the case of Chicago. Although the noise can vary by time of day and day of the week and the map cannot determine noise levels in particular locations, such maps can show trends over time and help property owners insulate buildings from noise and help planners determine where to avoid putting outdoor public spaces or residential areas. The map, in other words, can help us take flight from where people are taking flights.

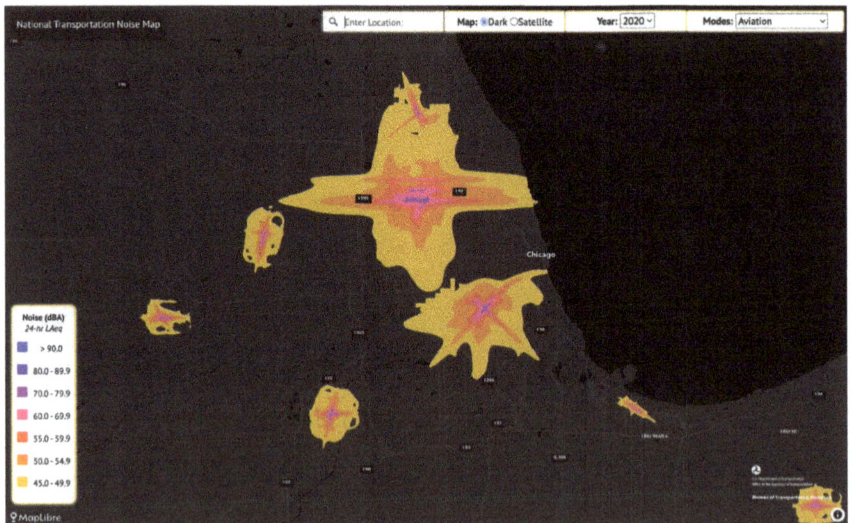

Transportation noise patterns relate to the locations of airports and other transportation nodes, such as those around Chicago. *Courtesy of the US Department of Transportation.*

Congestion

DISCIPLINE Public Health   **INDUSTRY** Transportation

ORGANIZATION Waze   **LOCATION** Washington, DC, USA

Geospatial tools have revolutionized the field of transportation planning through volunteered geographic information (VGI), which lets people contribute real-time information about traffic and road conditions and share it widely. The VGI platform Waze lets users upload information about traffic conditions and road closures so that others can avoid the delays that usually accompany such situations. As this case study of Washington, DC, during recent protests shows, Waze has value in helping commuters avoid traffic congestion and guiding people to the location of certain events. The app shows traffic jams, with icons that indicate the location of reported closures, the paths of garbage trucks and snowplows, and where other "Wazers" are situated. It reminds us that a map is only as good as the people who make it and is maybe the best when people help make it themselves.

Water examples

Water utility

DISCIPLINE Public Policy   **INDUSTRY** Water

ORGANIZATION Evansville Water and Sewer Utility   **LOCATION** Evansville, Indiana, USA

When it comes to delivering water to customers, accuracy and precision matter. The Evansville Water and Sewer Utility used global navigation satellite systems (GNSS) to update their GIS maps to provide centimeter-level accuracy to the locations of their service lines. The utility staff uploaded their data to ArcGIS Online, which made it immediately accessible, and created dashboards that let staff analyze and visualize the information coming from the field. The utility also developed a web app that lets others in the city access as-built drawings, historical maps, and customer service requests. This has enabled the utility not only to respond to service problems in a quick and accurate way but to manage its assets more efficiently and effectively. It's a workflow as fluid as the flow of water itself.

Oceanography

DISCIPLINE Oceanography **INDUSTRY** Science

ORGANIZATION Zooniverse **LOCATION** Global

Oceanographers know the hurdles that exist in mapping marine life, in part because of the sheer expanse and volume of the oceans. The nonprofit organization Zooniverse has taken that on as one of its many citizen science projects. A partnership between Chicago's Adler Planetarium and the universities of Oxford and Minnesota, Zooniverse is a "platform for people-powered research," with millions of volunteer participants aiding professional researchers, much of this effort involving mapping. The Click-a-Coral project exemplifies its work. It asks volunteers to draw boxes around different types of coral captured in deep-sea videos to train artificial intelligence (AI) to identify species, with the goal of automating the identification process in the future. The AI will eventually be able to recognize coral species and provide useful information more quickly than researchers could ever do manually, speeding data analysis and protection efforts. Ironically, there may be no tool more important to the conservation of the natural environment than AI.

Through the Click-a-Coral project, artificial intelligence will one day be able to recognize various coral species in deep-sea videos and photos. *Courtesy of Zooniverse and the Click-a-Coral project on Zooniverse.*

Notes

1. The Rocket Factory, artwork by Marcus Neustetter. Image courtesy of the artist.
2. Esri Nederland and Community Maps Contributors.
3. Liping Di and Weiguo Han, Center for Spatial Information Science and Systems, George Mason University. Zhengwei Yang and Rick Mueller, USDA National Agricultural Statistics Service.
4. Hengl, T., J. Mendes de Jesus, G. B. M. Heuvelink, M. Ruiperez Gonzalez, M. Kilibarda, et al. (2017). "SoilGrids250m: Global Gridded Soil Information Based on Machine Learning." PLoS ONE 12 (2): e0169748. doi:10.1371/journal.pone.0169748.
5. U.S. Energy Information Administration (January 2026).
6. This fictional dataset includes administrative, smartphone, and vehicle records, which were created specifically for use in movement and network analysis tutorials using ArcGIS AllSource.
7. © what3words Limited. All rights reserved.
8. US Department of Transportation, Bureau of Transportation Statistics. August 20, 2025. National Transportation Noise Map. Accessed December 22, 2025. https://maps.dot.gov /BTS/NationalTransportationNoiseMap/.

References

Agriculture: https://nassgeodata.gmu.edu/CropScape/

Agronomy: https://soilgrids.org/

Architecture: https://marcusneustetter.com/rocket-factory/

Austin ISD: https://www.austinisd.org

Biodiversity: https://storymaps.arcgis.com/stories/01c17e62ab224a7d9ba1e5dc3b23438e

Cables: https://richiecarmichael.github.io/cable/index.html

Climate change: https://storymaps.arcgis.com/collections /ad2309f0fa924f94851e2f03fa5f122a?item=6

Congestion: https://www.waze.com/live-map

Consumer behavior: https://gisgeography.com/huff-gravity-model/

Criminology: https://learn.arcgis.com/en/projects/find-the-meeting-locations-of-a-network-of-associates/

Defense: https://www.esri.com/news/arcuser/1010/3danalysis.html

Demographics: https://www.esri.com/news/arcuser/0311/population-drift.html

Dentistry: https://iowa.maps.arcgis.com/apps/dashboards /28a85e1c5a3e414fb3102e26f3fac98c

Digital equity: https://geoc.umn.edu/projects

Ecology: http://www.refractions.net/expertise/casestudies/2007-11-hectares-bc/

Economics: https://data.worldbank.org/

Electrical grid: https://www.arcgis.com/apps/View/index.html?appid= 390c4b30b92b46a6a8477b5ec3598334

Esri: https://www.esri.com

Equity: https://storymaps.arcgis.com/stories/dd2fc8c8c4904ce9b1f2dd8661b9d6ec

Evansville Water and Sewer Utility: https://ewsu.com

Geography: https://what3words.com/clip.apples.leap

GIS Geography: https://gisgeography.com/

Intelligence: https://www.jasondavies.com/maps/voronoi/airports/

Iowa Department of Public Health: https://hhs.iowa.gov/

ISRIC: https://isric.org/

Jason Davies: https://www.jasondavies.com/

JLL: https://www.jll.com

Landscape architecture: https://storymaps.arcgis.com/stories
    /4f4a03c675e34f01b781fd00a4b72cad

Marcus Neustetter: https://marcusneustetter.com/

Noise: https://maps.dot.gov/BTS/NationalTransportationNoiseMap/

Oceanography: https://www.zooniverse.org/projects/jordan-pierce/click-a-coral/talk
    /subjects/109126271

Oregon DLDC: https://www.oregon.gov/lcd/pages/index.aspx

Pipelines: https://www.arcgis.com/home/item.html?id=
    c745d9f4b81e42f3a54aee7aaa396975

Physical safety: https://www.youtube.com/watch?v=bv1nzhaUZKk

Public health: https://data.who.int/dashboards/covid19/cases

Real estate: https://www.esri.com/about/newsroom/publications/wherenext/jll-self-serve-
    and-full-serve-analytics

Refractions Research: http://www.refractions.net/

Satellites: https://richiecarmichael.github.io/sat/index.html

Seismology: https://www.arcgis.com/apps/View/index.html?appid=
    77990f072741415ea47353db2b00ab1a

Sustainable Development Goals: https://www.cehd.umn.edu/global/sdg

Trubel & Co.: https://www.trubel.co

University of Minnesota: https://umn.edu

US Department of Transportation: https://www.transportation.gov/

USDA NASS: https://www.nass.usda.gov/

Water utility: https://ewsu.com

Waze: https://www.waze.com/

what3words: https://what3words.com/about

World Bank: https://www.worldbank.org

World Health Organization: https://www.who.int/

Zooniverse: https://www.zooniverse.org/

CHAPTER 7

Building a geospatial support center

This chapter looks at what it takes to have GIS thrive at your university. There isn't one right way to do this. That fact is part of the point of this book: Every campus is different—different culture, different structure, different needs—so the approach has to fit your environment. The "Advice for License Administrators" section in this chapter looks at overcoming some of the biggest obstacles, such as licensing and managing access to software and applications. Many universities struggle with how to make software and data available to everyone, not just a few departments or individuals who happen to have or historically have had access. We'll look at some of the different models for enabling GIS at your university. What's important is understanding what the key pieces are and how they may apply in your own setting. Understanding what gets in the way and how to eliminate those roadblocks is key to creating a campus where GIS can thrive.

We hope the first six chapters of this book have piqued your interest in building or expanding a geospatial support center to serve your institution. Let's dive into more detail on how to do this by looking at what services to offer, models for staffing and funding, advice for Esri's ArcGIS license administrator, and additional resources.

GIS support center models

There are many ways to approach a GIS support center on your campus. We see the functions of the support center as distinct from teaching GIS classes, but on many campuses the functions will be delivered by the same department or even the same person. A number of models exist, all with good examples from institutions that have successfully implemented a successful support center. This table lists the models and their respective advantages so that you can decide which would work best for your campus.

Models for GIS support centers

	Libraries	Information Technology	Dedicated GIS Centers	Departments or Individuals	Community Efforts
Mission to support all students, faculty, and staff	X	X	X		X
Expertise across all disciplines	X				X
Repository for data, maps, and other resources	X				X
Requires hiring or training GIS professionals	X	X			
Reliance on central funding	X	X			
More entrepreneurial; less reliance on central funding			X		X
May face challenges supporting colleagues outside the unit				X	

Libraries

A lot has been written about how libraries can be an excellent home for geospatial support. David Cowen, professor emeritus at the University of South Carolina, created a story map guide on how university libraries can provide GIS services. Dr. Cowen notes that libraries are central units with a mission to serve all students, faculty, and staff across the institution. Libraries have resources that are especially helpful for those who are new to geospatial work, and those resources include access to spatial data and computer labs. Additionally, libraries generally have the data and expertise to support spatial fluency. His story map provides summaries and links to GIS centers at 20 colleges and universities. Institutions with good examples of library-based support centers include Arizona State University, Clemson University, and Bucknell University.

Information technology

Some campuses find their office of information technology (IT) to be a suitable home for GIS support services. Like libraries, being part of a central unit allows for support across the entire campus or system of campuses. On many levels, GIS can be seen as another form of technology and tools that everyone uses (for example, Google, Microsoft, and so on)—certainly when approaching the administration of and need for computing resources needed to host GIS software and services. The IT department will have an established help desk ready to take questions, seasoned experts to manage user access, and academic technologists whose job is to work with instructors and learners in the classroom. An investment in staff is needed to teach GIS and the importance of spatial thinking.

Good examples of GIS centers based in the IT office include New York University and Swarthmore College.

Dedicated GIS centers

The geospatial model with U-Spatial at the University of Minnesota, as described in chapter 1, is a central unit model. In our case, it's based in the university's Research and Innovation Office—but it could also be situated in other central units.

Examples of dedicated GIS centers include the Yale Center for Geospatial Solutions, Geospatial Centroid at Colorado State University, and U-Spatial.

Departments or individuals

Some schools that have a geography department may serve as the de facto provider of geospatial support. The primary concern with this approach is that the department or individual may not have the bandwidth to provide support outside normal job duties. A little success and adoption of GIS can quickly overwhelm a department with good intentions.

At the University of Michigan, Peter Knoop drives the support of GIS through the IT infrastructure in the College of Literature, Science, and the Arts in collaboration with the libraries. In some cases, individual faculty champions take the lead in pushing geospatial initiatives forward.

Community efforts

This approach probably has the most potential to spread GIS support across your campus. It's based on the realization that a single center would struggle to provide all the expertise and resources needed to support geospatial services across an institution. Let's use U-Spatial as an example.

U-Spatial isn't as big as it looks. We're often credited with providing services that other departments and centers take the lead on. We don't fund ArcGIS software licenses—that's the responsibility of the university's Office of Information Technology (OIT). For more than a decade, the university has been part of a consortium ArcGIS license with Minnesota State, a system with 54 campuses incorporating technical, community, and large universities. OIT also manages the computer labs and installs GIS software on the computers and virtual machines that our colleagues use.

Departments such as Geography, Environment and Society, Computer Science, Forest Resources, Public Policy, Public Health, and others teach GIS courses, from introductory classes to advanced topics, such as machine learning, AI, and advanced analytics. U-Spatial is part of an academic unit, but it doesn't teach credit-bearing courses, although U-Spatial has taught several skills-based workshops since its inception.

The libraries are the primary provider of data services, including discovery, procurement, data management plans, and archiving and preservation. The Big Ten academic Alliance Geoportal, which is hosted in the libraries, provides data discovery and access from nearly all the participating institutions. The libraries also support instructors through course-specific web portals and many "LibGuides" (library guides).

The Geospatial Information Science Student Organization (GISSO) brings students together with an annual career and networking fair, which welcomes the GIS community to meet young professionals. IPUMS, which provides census and population data from around the world, and the Polar Geospatial Center, which maps both polar regions, are located at the University of Minnesota, giving us easy access to experts in this field. This list isn't exhaustive, and we always cringe a little when making a list because someone will be left off.

No single center can provide all these support functions, and someone (maybe you) should take the lead on pulling people and resources together to prevent gaps or wasteful duplication of services. This lead role can be one person, a committee, a council—whatever makes sense for how things work on your campus. When U-Spatial was formed in 2011, the mailing list of GIS experts consisted of 87 colleagues; that list is now more than 2,000 people. When you look across your institution, you'll find partners in building a support network.

Advice for GIS support center adoption

Changing the mindset of campus leadership probably sounds hard and a little daunting, right? Most people going into the GIS profession are more interested in the tools, technology, and data and are attracted to solving problems because that's what allows us to answer the big questions of the day. Not everybody sees the power of thinking spatially the first time they open GIS software or make a map. GIS professionals need to improve their storytelling about geospatial and how the power of GIS can change the world. We have amazing stories to tell and the tools to tell a story. We need to learn how to tell these stories or partner with storytellers who can do so for us. We need to adapt the story to fit the specific audience; a story map is a good tool for teaching someone about a topic, but a college president may not have the time to digest a story map.

Thinking about how we communicate the value and importance of GIS reminds us of an exercise we've been doing with our incoming masters of GIS students for more than a decade, using the StrengthsFinder assessment from Gallup (a common measure for helping people understand where their strengths lie across areas such as executing, influencing, relationship building, and strategic thinking).

What we've found consistently over more than 10 years is that students entering our program are exceptional problem solvers. They're strong at learning tools, mastering technology, and building relationships with one another. Where they tend to be weaker is in the influencing category, particularly with communication and maximizing impact. These are critically important strengths when you're trying to make the case for GIS to senior leadership, especially a university or college president. If those strengths aren't naturally present, we need to work on them deliberately or partner with others who can help fill those gaps.

That brings us to the importance of BLUF—bottom line, up front. When you're communicating with senior executives, you don't have the luxury of long explanations or detailed justifications. You can't spend the time trying to convince them step by step. The key takeaway has to be front and center. What do they need to know, and why does it matter right now? And finally, what action are you recommending and what resources will be needed to achieve results?

Story maps are a powerful way to tell a story and build an argument, but they may take too much time for a busy executive to digest. In many cases, what's more effective is a half-page summary, something that fits into a PDF file, which can be printed, carried, and referred to during a packed day. That's quite different from asking someone to click through a web app or explore a detailed narrative. Those tools absolutely have their place, but audience matters, and how people consume information matters just as much as the content itself.

This kind of audience-aware communication is a skill set that many people in the GIS profession don't naturally have, and it's one that needs to be developed to build a network of GIS champions across your institution. Many people often want to help move GIS forward, and many ways are useful for doing so that don't require large financial investments. Resources are tight everywhere, and no institution ever gets all the funding it wants—or even needs—to do everything perfectly. Are there key faculty that can be a proponent of GIS at your institution?

Finally, it's worth thinking creatively about how GIS support is funded on campus. At U-Spatial, we do a significant amount of consulting work, and we've structured things so that much of the profit and indirect costs from grants flow back to U-Spatial. That revenue supports services for the common good, such as a help desk, training workshops, and events. It's been a sustainable model for nearly a decade.

Is something like this possible at your institution? What would it look like in your context? Maybe your college or university already uses chargebacks to fund the ArcGIS license (a chargeback is a reversal of a credit or debit transaction initiated by a bank). If so, could modest contributions from multiple departments be pooled to support a geospatial center to fully realize the value of the platform? Being entrepreneurial and finding ways to fund GIS support reduces the dependency on limited and uncertain central funding, which many other departments and centers are competing for.

Advice for license administrators

We want to offer advice about and insight into the lessons you've learned through starting a GIS support center. The community of people actively supporting GIS across many institutions is growing and generous with their time and expertise. This includes what we've learned from countless conversations with colleagues at the annual Esri Education Summit, Esri Developer and Technology Summit, and interactions online. The following list provides quick points on what's needed for effectively managing software licensing and other matters at a GIS center, as well as ideas on how to demonstrate the value of geospatial capabilities to your campus leadership.

- Turn on single sign-in so that all students, faculty, and staff can easily access ArcGIS Online. This topic comes up frequently at conferences, and for good reason. It's one of the most effective ways to reduce administrative overhead and is absolutely worth the technical and policy changes needed to make it happen.

- Our recommendation is to treat GIS resources as a common-good resource, rather than using a chargeback model that assigns a portion of the cost of an ArcGIS license to the individual departments or users of the tools. In our experience, chargebacks have created more administrative work than they're worth, and there hasn't been a meaningful cost-benefit return from trying to recover small amounts of funding that way. When we made an ArcGIS license a common-good resource, we saw a significant increase in the use of the tools and new opportunities for classroom use, especially in courses that are not specifically about GIS. This aligned with the rise of web-based GIS and tools such as ArcGIS Online. Making these resources freely available to all students, faculty, and staff has been essential for promoting spatial fluency across the university. At the University of Minnesota, the Office of Information Technology pays for the annual ArcGIS license. Every few years, we are asked to justify the cost of the license, and we can quickly show the value, in terms of ArcGIS Online credits and access to software, which far outweighs the cost to the university. This is what works for Minnesota; each college and university should think about its own culture and what approach would be the best fit. It really does vary by institution, but you should be able to find information in this book to make the case for making GIS software free for all to use.

- Many blogs and user groups are available that can help with best practices for administering ArcGIS Online subscriptions and other Esri infrastructure. Within those blogs is a wealth of information on how to manage credits and users, how to use scripting to automate administrative tasks, and general best practices for managing your environment. In addition to the blogs, a community of licensed administrators are willing to answer questions and

share their experiences. The Esri Community resources are an excellent place
to connect with others and learn from their approaches.

- Find your partners who can think and thrive outside the box with you. People
  drawn to GIS are problem solvers who sometimes prefer or need to stretch
  the norms of an organization to get things done. A geospatial support center
  doesn't teach GIS courses for credit; the purpose of the support center is to
  provide infrastructure so that instructors and students can access software,
  tools, and apps and focus on teaching and learning GIS.

Resources

As we worked on this book, we constantly discovered amazing work being done in academia (and beyond) around the world. This list of resources can complement your own
discovery efforts.

Organizations

- **American Association of Geographers (AAG)**: https://www.aag.org. A global
  organization that seeks to use geography to understand big problems by
  building communities.
- **Esri Higher Education**: https://www.esri.com/en-us/industries/higher-
  education. Your starting point to learn more about bringing GIS to your campus.
- **Esri Innovation Program (EIP)**: https://www.esri.com/en-us/c/industry/
  education/20/esri-innovation-program. This program's goal is to align
  with industry trends and workforce needs to motivate students to advance
  geospatial science in research and education. Each campus names a student
  of the year to be in the running for Esri Student of the Year, with a free trip to
  the Esri User Conference.
- **Geospatial World**: https://geospatialworld.net. A resource for tracking
  geospatial efforts and their intersection with academia.
- **GIS Certification Institute**: https://www.gisci.org. The GIS Professional (GISP)
  designation is a portfolio- and exam-based certification demonstrating an
  individual's GIS expertise.
- **Open Geospatial Consortium (OGC)**: https://www.ogc.org. This consortium
  is known for its leadership in collaboratively developing standards to drive
  solutions across the globe in all sectors.
- **University Consortium for Geospatial Information Science (UCGIS)**:
  https://www.ucgis.org. The professional hub for the academic GIS community,
  primarily based in the United States.
- **YouthMappers**: https://www.youthmappers.org. With more than 400 university
  chapters in 80 countries, YouthMappers introduces young leaders to mapping
  and GIS to create resilient communities.

Additional information

- **"7 Essential Skills Every GIS Manager Needs"**: https://www.esri.com/about/newsroom/arcuser/7-essential-skills-every-gis-manager-needs. A blog post about skills needed for future GIS professionals and managers.
- **ArcGIS Governance Best Practices for Education Customers**: https://community.esri.com/t5/education-blog/arcgis-governance-best-practices-for-education/ba-p/1646206. A series of Esri blogs, webinars, and workshops filled with tips on how to manage ArcGIS Online and other Esri resources.
- **Big Ten Academic Alliance Geospatial Information Geoportal**: https://geo.btaa.org. A helpful resource for data discovery and access that pulls together resources from 17 institutions in one easy-to-use application.
- ***Does My University Have ArcGIS?***: https://experience.arcgis.com/experience/bd3d8f3db9f942d19d08476b1ce23092. A story map managed by Esri that shows which campuses currently have ArcGIS licenses.
- **Gallup StrengthsFinder (CliftonStrengths)**: https://www.gallup.com/cliftonstrengths. A tool for exploring your talents and identifying how they can accelerate your professional development.
- **Geospatial Knowledge Infrastructure (GKI)**: https://geospatialworld.net/gki. Whereas the NSDI is US-focused, the GKI is global. It outlines a plan for a robust geospatial infrastructure and workforce.
- **GIS&T Body of Knowledge**: https://www.ucgis.org/site/gis-t-body-of-knowledge. A living textbook on GIS concepts, including data management, applications, computing platforms, analysis and modeling, cartography, and societal impacts.
- **GISGeography**: https://gisgeography.com/spatial-analysis-periodic-table. The periodic table of chemical elements, reimagined for all aspects of spatial analysis.
- **Guide to the Geographic Approach**: https://www.guidetothegeographicapproach.com. Developed at the University of California, Santa Barbara, this comprehensive resource provides lessons and tools for teaching GIS and spatial thinking.
- **Harvard Summer Institute**: https://gis.harvard.edu/gis-institute. Long-running, comprehensive GIS workshop that teaches numerous concepts in two weeks.
- **Institute for Geospatial Understanding Through an Integrative Discovery Environment (I-GUIDE)**: https://i-guide.io. An NSF-funded initiative that brings people from many fields together to use geospatial data and tools to solve real-world sustainability and resilience problems, with an emphasis on open collaboration, education, and broader participation in research and innovation.

- **"Lehigh University's Teacher Education Program Illustrates Why GIS Is a Powerful Classroom Tool"**: https://www.esri.com/en-us/lg/industry /education/stories/lehigh-universitys-teacher-education-program-illustrates-gis-powerful-classroom-tool. A user story whose title says it all!
- **National Spatial Data Infrastructure (NSDI)**: https://www.fgdc.gov/nsdi-plan. A broad plan describing how all levels of government and other sectors can collaborate on building a geospatial infrastructure that widely benefits through governance, people, and data and technology.
- ***University Libraries as Providers of GIS Services: A Guide***: https://storymaps. arcgis.com/collections/a17f0c19c5b94bdbaefbe6001c2dd923. A story map by David Cowen outlining several institutions with good GIS support services.
- **Yale Accelerators**: https://geospatial.yale.edu/learning. Two five-day programs that focus on GIS and remote sensing.

About Esri Press

Esri Press is an American book publisher and part of Esri, the global leader in geographic information system (GIS) software, location intelligence, and mapping. Since 1969, Esri has supported customers with geographic science and geospatial analytics, what we call The Science of Where. We take a geographic approach to problem-solving, brought to life by modern GIS technology, and are committed to using science and technology to build a sustainable world.

At Esri Press, our mission is to inform, inspire, and teach professionals, students, educators, and the public about GIS by developing print and digital publications. Our goal is to increase the adoption of ArcGIS and to support the vision and brand of Esri. We strive to be the leader in publishing great GIS books, and we are dedicated to improving the work and lives of our global community of users, authors, and colleagues.

Acquisitions
Stacy Krieg
Alycia Tornetta
Jenefer Shute
Katie Gezi

Product Engineering
Craig Carpenter
Maryam Mafuri

Editorial
Carolyn Schatz
Mark Henry
David Oberman

Production
Monica McGregor
Victoria Roberts

Sales & Marketing
Eric Kettunen
Sasha Gallardo
Beth Bauler

Contributors
Christian Harder
Matt Artz

Business
Catherine Ortiz
Jon Carter
Jason Childs

Related titles

The Power of Where

Jack Dangermond, Esri, and
the GIS Community

9781589486065

Exploring GeoAI

Ismael Chivite, Nicholas Giner,
and Craig Carpenter

9781589489080

Getting to Know ArcGIS Pro 3.6

Michael Law and Amy Collins

9781589488984

Map Use, volume 1, ninth edition

Aileen R. Buckley, A. Jon Kimerling, and
Patrick J. Kennelly

9781589487758

For more information about Esri Press books and resources,
or to sign up for our newsletter, visit

esripress.com.